U.S. Marine Corps Combat Water Survival

Plus

USMC Land Navigation Sergeants Course

Department of Defense

Pentagon
Publishing

MCRP 3-02C

Marine Combat Water Survival

U.S. Marine Corps

PCN 144 000069 00

To Our Readers

Changes: Readers of this publication are encouraged to submit suggestions and changes that will improve it. Recommendations may be sent directly to Commanding General, Marine Corps Combat Development Command, Doctrine Division (C 42), 3300 Russell Road, Suite 318A, Quantico, VA 22134-5021 or by fax to 703-784-2917 (DSN 278-2917) or by E-mail to **morgann@mccdc.usmc.mil**. Recommendations should include the following information:

- Location of change
  - Publication number and title
  - Current page number
  - Paragraph number (if applicable)
  - Line number
  - Figure or table number (if applicable)
- Nature of change
  - Add, delete
  - Proposed new text, preferably double-spaced and typewritten
- Justification and/or source of change

Additional copies: A printed copy of this publication may be obtained from Marine Corps Logistics Base, Albany, GA 31704-5001, by following the instructions in MCBul 5600, *Marine Corps Doctrinal Publications Status*. An electronic copy may be obtained from the Doctrine Division, MCCDC, world wide web home page which is found at the following universal reference locator: **http://www.doctrine.usmc.mil**.

Unless otherwise stated, whenever the masculine gender is used, both men and women are included.

DEPARTMENT OF THE NAVY
Headquarters United States Marine Corps
Washington, D.C. 20380-1775

6 January 2003

FOREWORD

Marine Corps Reference Publication (MCRP) 3-02C, *Marine Combat Water Survival*, provides Marine Corps combat water survival techniques, procedures, and training standards. This publication also teaches Marines to cross water obstacles and perform water rescues correctly and safely.

This publication is the foundation for teaching Marines correct water survival techniques and procedures that are used throughout the Marine combat water survival program (MCWSP). Once an individual or a unit has completed the MCWSP, this publication can be used as a refresher course before water operations.

MCRP 3-02C supersedes Fleet Marine Force Reference Publication (FMFRP) 0-13, *Marine Combat Water Survival*, dated 16 September 1991.

Reviewed and approved this date.

BY DIRECTION OF THE COMMANDANT OF THE MARINE CORPS

EDWARD HANLON, JR.
Lieutenant General, U.S. Marine Corps
Commanding General
Marine Corps Combat Development Command

Publication Control Number: 144 000069 00

MCRP 3-02C Marine Combat Water Survival

Table of Contents

Chapter 1. Survival at Sea

Abandoning Ship 1-2
Jettisoning Equipment 1-2
Abandoning Ship Technique 1-3
Modified Abandoning Ship Technique 1-6
Surface Burning Oil Swim 1-9
Surviving With a Pack 1-11
Preparing Equipment 1-12
Tying Waterproof/Plastic Bags 1-13
Packing the Pack 1-13
Swimming With the Pack 1-13
Staying Afloat With a Life Preserver 1-15
Inherently Buoyant Life Preservers 1-15
Inflatable Life Preservers 1-15
Staying Afloat Without a Life Preserver 1-18
Floating With an Inflated Blouse 1-18
Floating With Inflated Trousers 1-20
Sling Method 1-20
Splash Method 1-26
Blow Method 1-32
Avoiding Heat Loss in Cold Water 1-36
Individual Protection From the Cold 1-37
Group Protection From the Cold 1-38
Drownproofing Methods 1-39
Crawl Stroke 1-40
T-Method 1-44
The Sweep 1-46

Breast Stroke....1-48
Side Stroke....1-52
Elementary Backstroke....1-57

Chapter 2. Water Rescues

Reaching Rescue Techniques....2-2
Reach....2-2
Reach From a Deck....2-3
Arm Extension....2-3
Leg Extension....2-4
Wading Assist....2-4
Throw....2-5
Lifesaving Approaches....2-7
Front Surface Approach....2-7
Rear Approach....2-7
Approach Strokes....2-8
Crawl Stroke Approach Stroke....2-8
Breast Stroke Approach Stroke....2-11
Level Offs....2-13
Front Surface Approach....2-13
Single Armpit Level Off....2-14
Double Armpit Level Off....2-16
Rescue Techniques....2-18
Wrist Tow....2-18
Single Armpit Tow....2-19
Double Armpit Tow....2-21
Collar Tow....2-22
Cross-Chest Carry....2-22
Tired Swimmer's Assist....2-24
Defense Against a Drowning Victim....2-25
Block....2-25
Wrist-Grip Escape/Wrist-Grip Escape Alternative....2-26

Front Head-Hold Escape 2-27
Rear Head-Hold Escape 2-29
Administering First Aid/Rescue Breathing 2-31

Chapter 3. Treatment of Casualties and Avoidance of Dangerous Marine Life

Drowning 3-1
Symptoms 3-1
Treatment 3-2
Hypothermia 3-2
Symptoms 3-4
Survival Time 3-4
Treatment 3-4
Heat-Related Injuries 3-6
Burns 3-7
Common Medical Problems Associated with Sea Survival 3-8
Seasickness 3-8
Saltwater Sores 3-8
Blindness/Headache 3-9
Constipation 3-9
Sunburn and Dehydration 3-9
Dangerous Marine Life 3-10
Sharks and Barracudas 3-11
Sea Snakes 3-12
Poisonous Fish 3-12
Turtles and Moray Eels 3-12
Corals 3-12
Jellyfish, Portuguese Man-of-War, Anemones, and Others 3-12
Spiny Fish, Urchins, Stingrays, and Cone Shells 3-13

Chapter 4. Negotiating Water Obstacles

Tides....4-1
Surf....4-2
Plunging Waves....4-3
Spilling Waves....4-4
Surging Waves....4-4
Currents....4-4
Offshore Currents....4-4
Rip Currents....4-5
Littoral Currents....4-5
Back Bays....4-6
Rivers and Canals....4-6

Chapter 5. Fording Waterways

Selection of a Ford Site....5-1
Determine the Slope....5-2
Clinometer....5-2
Map....5-3
Line of Sight and Pace....5-3
Determine the Current Speed....5-4
Measure River Width....5-5
Calculate Downstream Drift....5-6
The Buddy System....5-7
Water Crossings....5-8
Care of Weapons....5-8
ISOMAT Raft....5-9
Poncho Raft....5-10
Construction of a Pack Raft....5-12
Single-Rope Bridge....5-13
High and Dry Crossings....5-17

Swift Current Crossings 5-18
Slow Current Crossings 5-19
Removal 5-19

Appendix. Knot Tying

Square Knot Appendix-1
Bowline Appendix-2
Hitches Appendix-2
Half Hitch Appendix-2
Two Half Hitches Appendix-3
Round Turn and Two Half Hitches Appendix-3
Butterfly Knot Appendix-4

Throughout history, water has posed special challenges to Marines and Sailors during times of both peace and war. Therefore, the inherent dangers associated with waterborne operations demand that Marines and Sailors receive proper water survival training. Combat units that are confident in their ability to work in and around water can use the water to their advantage in combat, and history is filled with examples of how the proper preparation or training for survival in water reduced or averted disaster.

USS *Indianapolis*

On Sunday, 29 July 29 1945, the heavy cruiser USS *Indianapolis* (the *Indy*) was en route to the Philippine Sea. Shortly before midnight (about 39 hours out of port), the *Indy*, running blacked out and unescorted, was rocked by two explosions on her starboard side. With communications smashed, the ship could not signal its distress and sank within 15 minutes.

Three life rafts and a floater net supported a few survivors, but the rest of the survivors drifted about, held up by rubber life belts or Mae Wests. About 60 seamen died the first night. Survivors assumed the ship would be reported overdue in Leyte, and that they would be rescued within 2 days. By Monday evening, however, panic began to set in as some life jackets lost their buoyancy from the long immersion. Some men even fought over life jackets, which resulted in at least 25 deaths. No one dared sleep for fear of losing his jacket.

Throughout the next several days, in-transit aircraft flew nearby without spotting the desperate seamen. As best they could, the men kept together, some tying long ropes to each other, floating like corks on a net.

Not until late Thursday morning, 3 1/2 days after the ship sank, were the men discovered. Luckily, a plane on a routine flight over the area spotted the survivors. When ships picked them up that night, the survivors learned they had never been reported overdue. Every one of the 1,196-man crew was a casualty; 880 crewmembers were listed as dead or missing. Although many lives were lost, the innovative and expedient use of flotation devices and float techniques employed by the survivors helped save hundreds of lives.

USS *America*

On Thanksgiving eve, 23 November 1995, the USS *America* made its way through the Arabian Sea. Twenty-year-old, Marine Lance Corporal (LCpl) Zachary Mayo was unable to sleep and, wanting some fresh air, made his way onto an open-air platform near the aircraft hangar bay, which was three levels below the sleeping quarters. While he was on the platform, the ship veered suddenly, throwing LCpl Mayo through the platform's protective bars and into the sea, 30 feet below.

Frantic, LCpl Mayo called out in vain to the watchmen on the flight deck, which was 64 feet above him. It soon became clear to him that the USS *America* would keep its course into the Gulf of Oman until his absence was discovered at morning muster. The LCpl took a moment to consider his situation. Since land was at least 100 miles away, swimming was suicide; he would have to stay afloat until a search party found him.

Using the techniques he had learned during combat water survival training, LCpl Mayo made a flotation device out of his coveralls and tried to relax. Meanwhile, business continued as usual aboard the USS *America*. Since LCpl Mayo was on special assignment with the hazardous materials division, his absence went unnoticed until a petty officer asked several Sailors if they had seen their shipmate recently. By the time a roll call had been completed, LCpl Mayo had been adrift at sea for over 24 hours. Although three, fixed-wing Viking aircraft were deployed to search for LCpl Mayo, most people aboard ship feared the worst.

After 34 hours at sea, LCpl Mayo was discovered by fishermen on a Pakistani fishing boat. LCpl Mayo's survival is a testament not only to his incredible physical courage, but also to the soundness of the lifesaving training and techniques he received during combat water survival training.

Chapter 1
Survival at Sea

As a Marine, you face a variety of potential water emergencies whenever you cross expanses of water: ships, watercraft, and amphibious assault vehicles (AAVs) can sink; aircraft can crash into the sea; or you can accidentally fall into the water. However, there are some basic precautionary measures you can take to protect your safety and reduce your chances of becoming a water casualty. Determine the following information as soon as you board any type of vessel. Your knowing the following information may save your life or the life of your fellow shipmates.

- How many life preservers and lifeboats/rafts are on board?
- Where are the life preservers and lifeboats/rafts located?
- What type of unit survival equipment is on board?
- Are individual survival kits issued to each person on board?
- How much food, water, and medicine do the survival kits contain? When was the last time the contents were inspected for proper quantities and shelf life expiration?
- Is there sufficient survival equipment available for the number of personnel?
- How many other personnel are there on board, and where are they located?
- What are the egress procedures for the ship, boat, watercraft, AAV, or aircraft?

Abandoning Ship

When you embark on a Navy ship, you will receive abandoning ship instructions from Navy personnel. If given the order to abandon ship, report to your designated assembly area and put on a life preserver. DO NOT inflate the life preserver until you are clear of the ship. Torn life preservers will not inflate and inflated life preservers can block you, and those behind you, from exiting the ship. A flotation device that has been inflated may also burst if you jump from a significant height. See pages 1-15 through 1-35 for staying afloat with and without a life preserver.

DO NOT remove your clothing, boots, or shoes before abandoning ship. Your trousers and blouse may be the only flotation devices available if your life preserver is faulty or becomes damaged, and your clothes can provide some insulation from the cold water. However, remove your soft cover and place it in a cargo pocket for later use. The soft cover is both lightweight and a good protection against sunburn caused by the sun's rays reflecting off the water.

Jettisoning Equipment

Equipment should be kept properly packed and waterproofed in case you have to abandon ship. If entering the water from a height greater than 30 feet, wearing your equipment (e.g., pack, helmet, gas mask) could cause injury. Upon impact with the water, the helmet will "cup" air inside of it. The chin strap may also create a "hanging effect" as you submerge from the force of the fall. Therefore, you should remove your helmet and gas mask before abandoning ship.

If you are unable to maintain buoyancy due to the amount of equipment secured to your pack and body, then jettisoning some of your equipment may become necessary. Equipment that you should always retain includes canteens of freshwater, first aid kit, soft cover, and survival kit. The survival kit should include first aid items, water purification tablets or drops, fire starting equipment, signalling items (e.g., flashlight, strobe light, chemlights), food procurement items, and shelter items. Other items in a survival kit include sunburn lotion and lip balm, knife, goggles/sunglasses, plastic bag, matches and lighter, mirror. See Marine Corps Reference Publication (MCRP) 3-02F, *Survival*, for a detailed list of survival items and applications.

Abandoning Ship Technique

When abandoning ship, safety considerations must be observed. Use the following technique when abandoning ship without your combat gear:

Place your hands on their opposite shoulders, forming a crisscross pattern.

Step to the edge of the ship's deck and check the water below for debris or survivors. If the water is clear, look straight ahead and prepare to jump. If the water is not clear, move to another location.

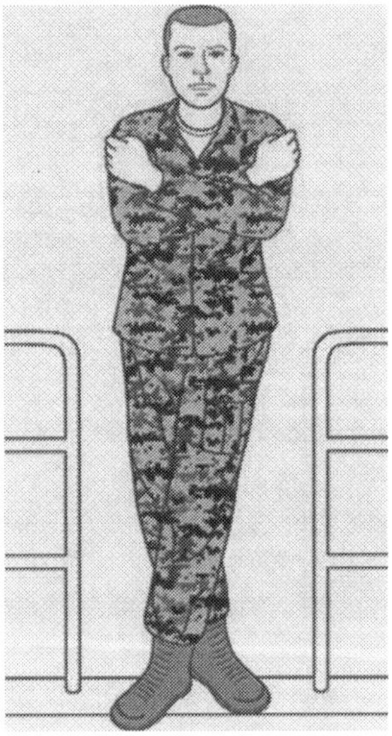

NOTE: DO NOT hold your nose as you abandon ship. If you do hold your nose, the force of impact into the water could jar your arm and hand and cause you to break your nose.

Step off the side of the ship with a smooth, 30-inch stride. DO NOT DIVE OFF THE SHIP. DO NOT LOOK DOWN AT THE WATER. LOOK STRAIGHT AHEAD. Looking down at the water can render you unconscious or cause injuries upon impact.

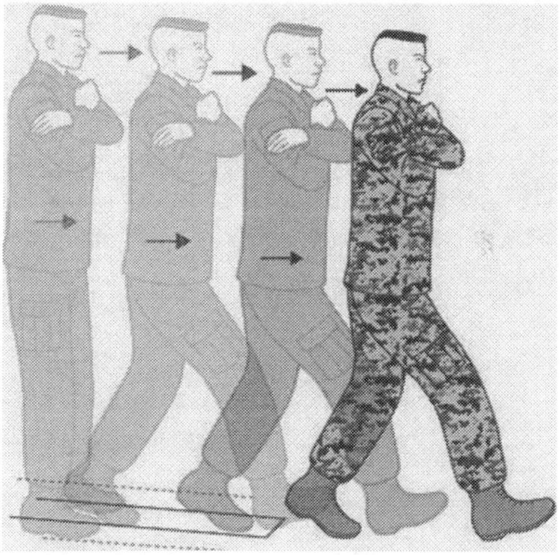

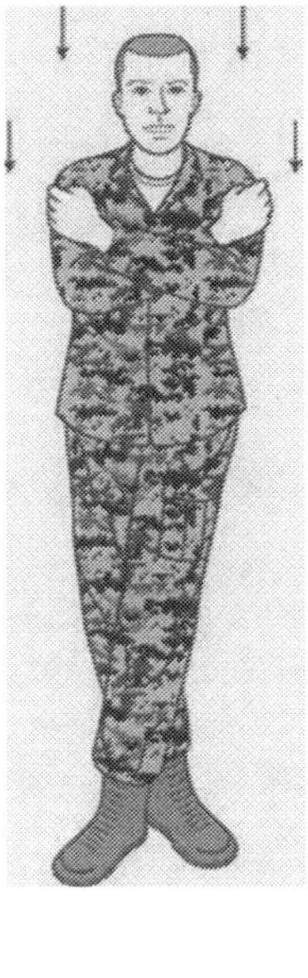

Bring your trailing leg forward during the fall. Cross your trailing leg behind your leading leg.

Keep your head parallel to the water's surface until hitting the water.

You should remain in the abandon ship position until your descent into the water has almost stopped. However, the weight imbalances in your body may cause you to be in a "J" shape under the water

.

Once your downward motion has ceased, your feet may be parallel with the ocean bottom or you may be nearly inverted with your feet over your head. To counteract potential disorientation, you should pause briefly and allow the natural buoyancy of your torso to bring your body to a nearly upright position.

Floating debris can cause hazards. Therefore, you should swim upward, extending one arm (hand is shaped as a fist) upward to feel for obstructions. If you encounter debris, try to push it away or surface in a different location.

Swim away from the ship. DO NOT LOOK BACK AT THE SHIP. Looking back could slow your movement away from the area. Remember, your objective is to leave the area as quickly as possible because—

- Equipment and debris may be falling from or spilling out of the ship.
- Additional casualties can occur if individuals abandoning the ship fall on top of swimmers already in the water.
- Swimmers close to the sinking ship may get pulled underneath the water by the suctioning effect of the ship as it goes under.

Modified Abandoning Ship Technique

When abandoning ship while wearing full combat gear (weapon, helmet, and a properly waterproofed pack), safety considerations must be observed. The modified abandoning ship technique is used while wearing full combat gear and exiting from a height less than 30 feet.

WARNING

When exiting from a height greater than 30 feet, remove your helmet and pack. Fasten the helmet to the pack or place it inside the pack before jettisoning. If jettisoning gear from the ship or aircraft, check the water below for survivors before throwing the gear forward of the intended jump area. Once in the water, retrieve your gear and swim out of the area.

Place your weapon over one shoulder, muzzle down, with the weapon parallel to your side.

Place your arm and hand along the weapon and hold it to your side.

Take your free hand and place it on top of your helmet to prevent neck/spinal injury from the force of the water pulling upward on the helmet as your body enters the water.

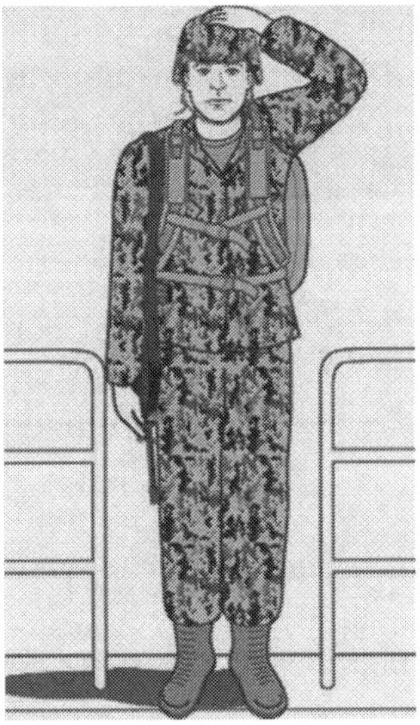

NOTE: An alternate method is to place your weapon over one shoulder, muzzle down, with the weapon parallel to your side. Reach across your body and grasp the sling of your weapon and hold it to your body. Take your free hand and place it on top of your helmet.

Step to the edge of the platform and check the water below for debris and survivors. If the water is clear, look straight ahead and prepare to jump. If the water is not clear, move to another location.

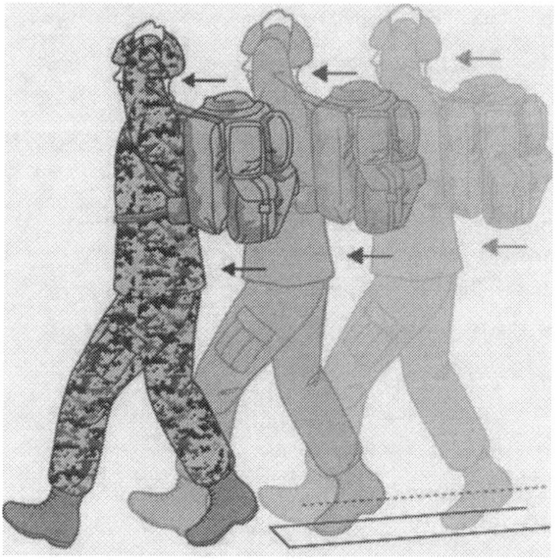

Step off the side of the platform with a smooth, 30-inch stride. DO NOT DIVE OFF THE SHIP, DO NOT LOOK DOWN AT THE WATER, LOOK STRAIGHT AHEAD. Looking down at the water can render you unconscious or cause injury upon impact.

Bring your trailing leg forward during the fall. Cross your trailing leg behind your leading leg.

Keep your head parallel to the water's surface until hitting the water.

You should remain in the modified abandoning ship position until your descent into the water has almost stopped. The buoyancy of a properly waterproofed pack will immediately pull you to the surface. Once you break the water's surface, unsling your weapon and loop the sling over your head. You may have to seesaw the sling so as to ensure the sling passes between your helmet and pack. Once over your head, loop the sling around your neck with the weapon aligned in the center of your body, muzzle down. Lean back on your pack and perform the combat travel stroke to exit the area.

> *Note:* You may also remove the pack (unsnap the quick-release strap on one side of the pack, and pass the strap to your other hand in order to maintain contact with the pack, which is functioning as your flotation device). Remain horizontal in the water with your pack under your chest. Perform a breast stroke kick.

Surface Burning Oil Swim

After you have abandoned ship, rise to the surface using the techniques shown on the page 1-6. However, you must remember that fuel from sinking ships or downed aircraft will float on the surface of the water. Therefore, you must move clear of the floating fuel by swimming away from the ship or aircraft as soon as possible. Either swim upwind (into the wind) of the ship/aircraft or swim against the current. Either method allows you to move away from the fuel and the wind/current will push the fuel past you. To properly execute a surface burning oil swim—

Extend your arms overhead as far as possible.

Wave your arms back and forth vigorously to splash a hole while moving upward.

Splash as long as possible to push burning fuel away from the surfacing area.

Use your arms and hands to sweep away fuel and debris.

Kick your legs in a constant breast stroke kick.

Extend your arms (palms outward) forward on the surface, arms shoulder-width apart.

Pull your hands in and back toward the chest.

Stop your hands in front of your face and rotate them so that your palms face forward (roughly halfway out of the water).

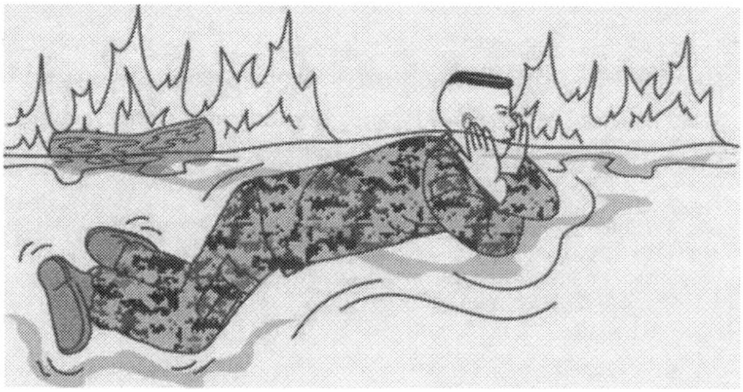

Sweep your arms forward to a full extension at the shoulder width. This splashes debris, oil, or burning liquids aside. To reduce the chance of fatigue, use two short splashes to the front to extend the path.

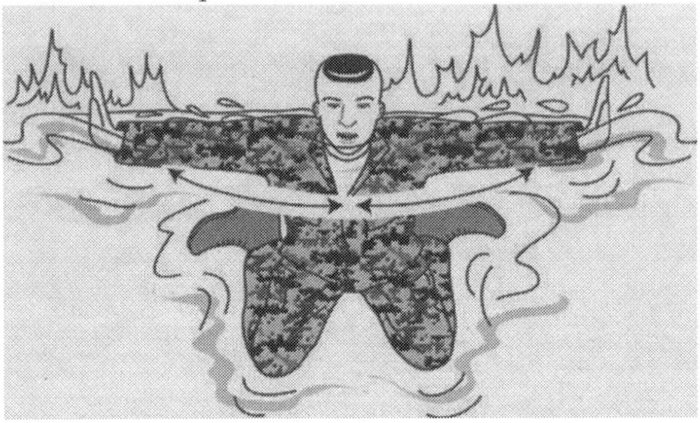

Repeat the preceding step as necessary while swimming clear of the area.

Surviving With a Pack

If packed properly, your pack will float, and it is your key piece of equipment for staying afloat and overcoming water obstacles. If the pack's contents are properly waterproofed, it can support you (with a combat load) in the water. Buoyed up by a waterproofed pack, you eventually emerge from the water with all your equipment (e.g., boots, helmet, flak jacket, weapon, survival items).

Your pack floats based on a scientific principle known as Archimedes' principle. This principle states that an object submerged in a liquid is buoyed up by a force equal to the weight of the liquid displaced (pushed aside) by the object. If the weight of the displaced liquid is greater than the weight of the object, the object floats. If the weight of the displaced liquid is less than the weight of the object, the object sinks. For example, a machine gun sinks in the water, but it still weighs less in the water than it does on land. Even though a machine gun sinks, it is still buoyed up by a force equal to the weight of the water it displaces. For this reason, you should not try to hold yourself or your equipment any higher out of the water than they would naturally float; doing so wastes both energy and body heat.

Preparing Equipment

Before packing your pack, you must prepare your gear/equipment. Tape or pad all sharp edges and equipment corners. Ideally, your gear/equipment is placed in plastic bags and the plastic bag is then placed inside the standard-issued, rubberized, waterproof bag. Waterproof bags are not completely water tight; henceforth, the added protection of first wrapping the gear/equipment in a plastic bag, then placing the plastic bag inside the waterproof bag. Large plastic bags (e.g., trash bags) work well for bulky equipment (e.g., sleeping bags, field jackets, shelter halves, gas masks). Small plastic bags work best for small items (e.g., shaving gear).

> *Note:* If a gas mask must be carried outside the pack, cover it with a waterproof bag.

You will need the following items, which are available through the supply system, to prepare your equipment for packing:

- Waterproof bags:
  Pistol bag, plastic size #1 (8" x 18").
  Rifle bag, plastic size #2 (10" x 56").
  Machine gun bag, plastic size #3 (15" x 56").
  Multipurpose bag, cover, plastic size #4 (20" x 84").

- Panama work vest (life preserver).
- Small plastic bag, self sealing (6" x 6" and 12" x 12").
- Riggers tape, olive drab (2 1/2").
- Willie peter bag, waterproof clothing.

Tying Waterproof/Plastic Bags

Try to remove the excess air from the waterproof/plastic bag before securing the bag's opening. If filled with air, the bag can burst if pressed from the outside. Perform the following steps to tie a waterproof/plastic bag:

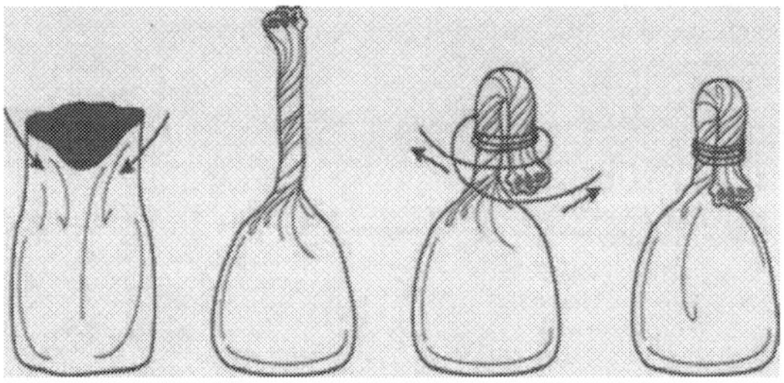

Packing the Pack

Place filled and tied bags inside the pack. Carefully handle sharp items (e.g., tent pegs, poles) to prevent puncturing the bag. Place items in the pack in order of expected use. Close the pack and its compartments. Attach sleeping mats or bags as high as possible on the outside of the pack.

Swimming With the Pack

A useful technique for propelling yourself forward while wearing a pack is the combat travel stroke. This technique allows you to float nearly horizontal and to propel yourself forward with bicycle-style kicks and breast stroke-style arm sweeps. To execute the combat travel stroke—

Body position
Keep the upper part of your body prone to the water, let your legs dangle horizontally, and keep your face up.

Arm action
Extend your hands out in front of your waist. Sweep your arms slightly downward and back 90 degrees to propel your body through the water. Move your hands back to the front of your waist. Repeat.

Leg action
Continuously move your legs in a bicycle-like movement, bringing your knees up high and step out.

Breathing
Keep your face out of the water during the stroke, and breathe freely.

Staying Afloat With a Life Preserver

The best form of flotation is to find any kind of floating object that will keep you and your equipment out of the water or minimize your exposure to the water. Life preservers are the best method, as they allow you to wear your clothes for heat retention and sunburn prevention. Marines use two basic classes of life preservers: inherently buoyant life preservers and inflatable life preservers.

Inherently Buoyant Life Preservers

Inherently buoyant life preservers are either vest-type (worn like a jacket) or yoke-type (worn around the neck). The preserver's outer envelope is either a cotton or water resistant material that encloses a removable fibrous glass or plastic foam filling.

In the Marine Corps, the most common type of inherently buoyant life preserver is the vest-type with collar, known as the kapock preserver. The kapock consists of collar straps, upper front chest straps, leg straps, and waist drawstrings that secure the preserver to you. The leg straps, which are fitted on both sides of the life preserver, ensure that the preserver remains around your chest while you are in the water. A chest strap is attached to the life preserver to facilitate lifting you out of the water. The strap can also be attached to other survivors or to lifeboats to reduce the fatigue that results from holding onto a floating/secured object by hand.

Inflatable Life Preservers

Marine Corps aircraft and AAVs have inflatable life preservers on board. The inflatable life preserver issued to Marines is known as the LPP (life preserver personal). The LPP is capable of both oral inflation and CO_2 cartridge inflation. The LPP consists of buoyancy chambers, CO_2 inflator, and an oral inflation tube. The buoyancy chamber is deck gray in color and is made from a neoprene-coated nylon fabric.

Inflatable life preservers must be stored in a cool, dry place. Heat, moisture, and light cause deterioration of the life preserver material. Do not stow CO_2 cylinders near steam lines or radiators. Heat can increase the pressure inside the cylinders causing them to explode. Avoid sharp edges in stowage. Sharp edges increase wear and tear on the life preservers and may also puncture inflatable buoyancy chambers.

> **CAUTION:** Do not inflate the life preserver until you are clear of the aircraft, ship, or vehicle. Torn life preservers will not inflate and inflated life preservers can block you, and those behind you, from exiting the aircraft, ship, or vehicle.

To don and adjust the LPP—

Remove the LPP from its storage container.

Fasten the belt fasteners in front with the pouch in the rear.

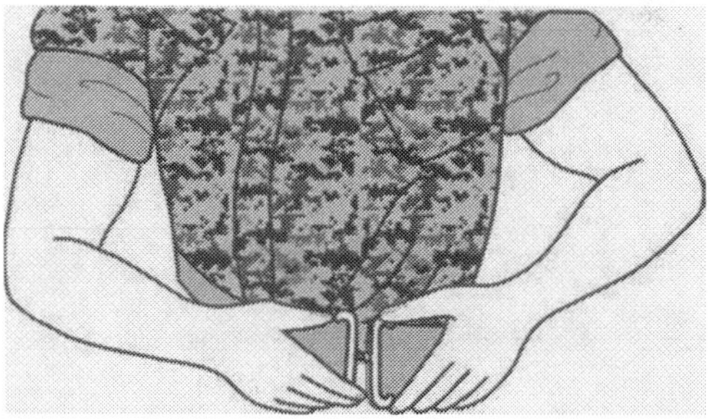

Adjust the belt to fit; secure any excess belt by mating the hook and pile tape.

Rotate the pouch to the front and re-adjust the belt if necessary.

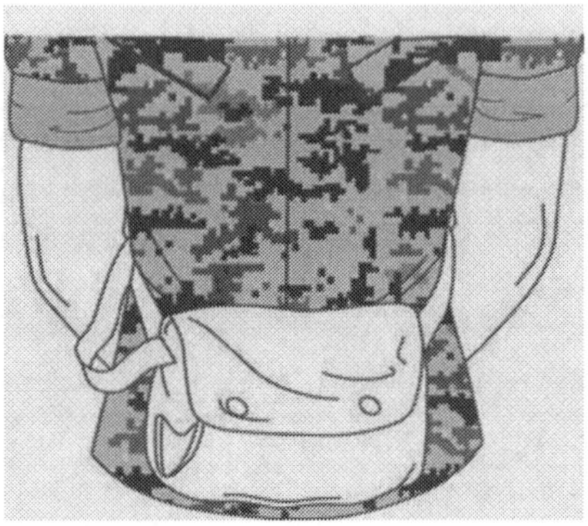

Open the snap fasteners on the pouch and unfold the life preserver.

Place the deflated life preserver over your head.

Place the storage container into the pouch after donning the life preserver.

Lift the lower end of the life preserver out of the pouch.

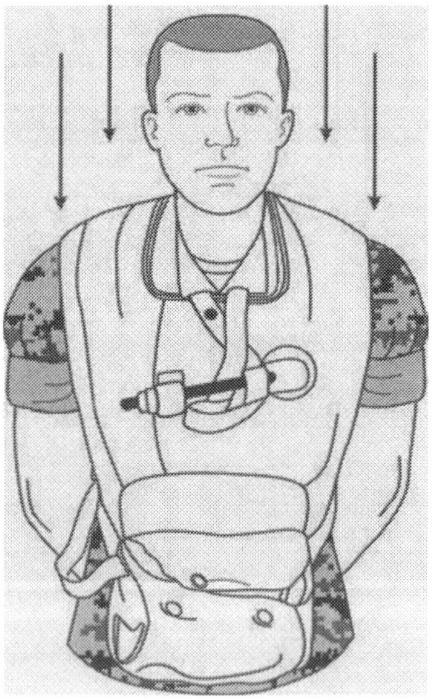

Inflate the life preserver by pulling on the lanyard attached to the CO_2 inflation valve or by blowing on the end of the oral inflation valve.

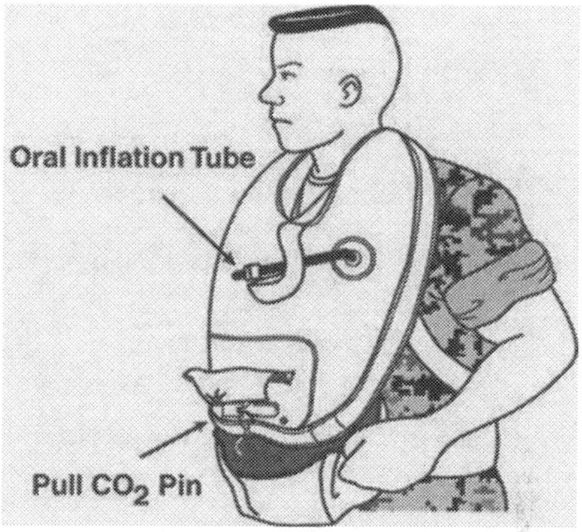

Staying Afloat Without a Life Preserver

You may be in open water without any floating objects or a life preserver to help you survive. If so, uniform blouses and trousers can be made into expedient flotation devices.

Floating With an Inflated Blouse

It is possible to float by a bubble of air trapped in the shoulders of your blouse. The air rises to the back and shoulders of the blouse and supports you at the water's surface. An inflated blouse is also a temporary flotation device used by weaker swimmers while trying to remove their trousers. There is a primary and an alternate way to create a bubble of trapped air in a blouse—

Primary Method
Turn the collar inside the blouse to help create a seal.

Unbutton top button and pull collar around mouth and nose.

Take a deep breath and bend forward slightly at the waist. Exhale one-half to three-quarters of a breath into the blouse.

Grasp and twist the collar with one hand to create a seal, this prevents air from escaping out from the collar.

Use your free hand and feet to stroke and kick to the surface.

Gather and hold the blouse tightly at the collar and stomach level to prevent the blouse from losing air if it floats up too high.

Splash water on the blouse periodically to prevent the material from drying, dry material allows air to escape.

Repeat inflation as required.

Alternate Method
Turn the collar inside the blouse to help create a seal.

Unbutton the second button from the top.

Take a deep breath and bend forward slightly at the waist.

Place your mouth and nose inside the hole created by the open button and exhale one-half to three-quarters of a breath into the blouse.

Grasp material at the unbutton portion and pull downward.

Use your free hand and feet to stroke and kick to the surface.

Splash water on the blouse periodically to prevent the material from drying, dry material allows air to escape.

Repeat inflation as required.

Floating With Inflated Trousers

In warm water, trousers can be used as a primary expedient flotation device. However, in cold water, submerging your head to remove and inflate your trousers results in heat and energy losses that negate the benefit of using the trousers as a flotation device. Once your trousers are inflated, you float motionlessly as if wearing a life preserver. If needed, assume the heat escape lessening posture (known as the HELP position, see page 1-37) to slow heat loss. As trousers dry, air leaks out of the legs. To slow this process, occasionally splash water on the fabric. Reinflate trousers as needed.

Sling Method

The sling method works if you are a strong swimmer or naturally very buoyant. Take the following steps to inflate trousers using the sling method:

Take a deep breath, bend over, and remove your boots.

NOTE: Retain your boots. Tie the boot laces together and suspend the boots from your blouse or hang them around your neck so that they rest on your chest.

Remove your trousers. Button or zip the trouser's fly closed. This allows you to control airflow.

Tie the bottoms of the trouser legs in a square knot. (The appendix illustrates various knots.)

Ensure that the front (fly) of the trousers faces you.

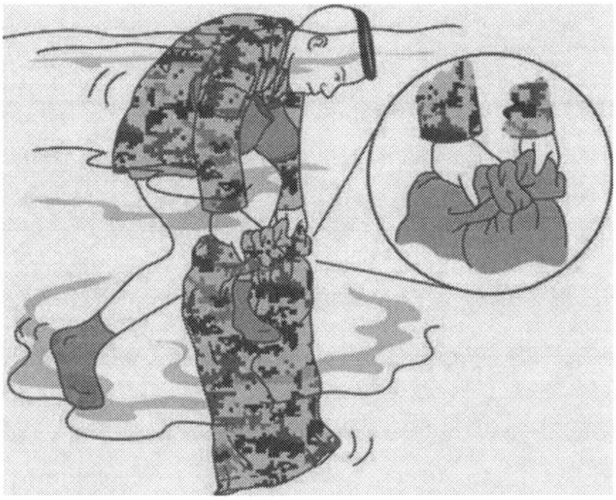

Hold the trousers above the water's surface and behind your head. Grasp both sides of the waistband and open with both hands.

Kick strongly to stay on top of the water while slinging the trousers overhead in order to trap air into them.

Once the waistband is submerged in the water, air is trapped in the legs.

Hold and seal the waistband underwater.

Slip the inflated legs over your head. Hold the waistband in toward your chest, the fly facing your body. To prevent air from escaping from the trousers, seal the waistband by either folding it or twisting it.

Lie back and relax, resting the back of your neck against the knot.

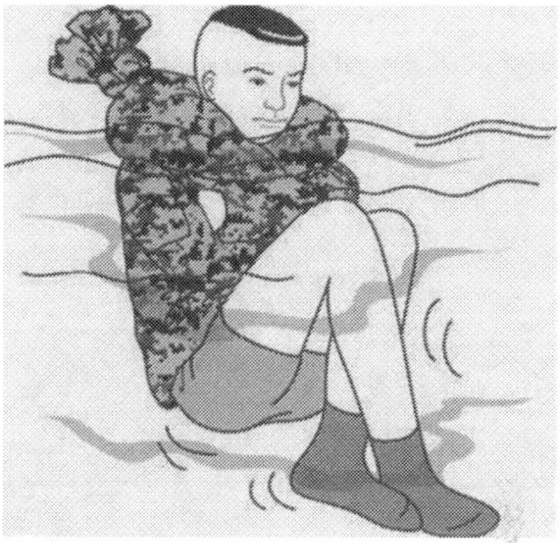

Splash water on the trousers periodically to prevent the material from drying. Dry material allows air to escape.

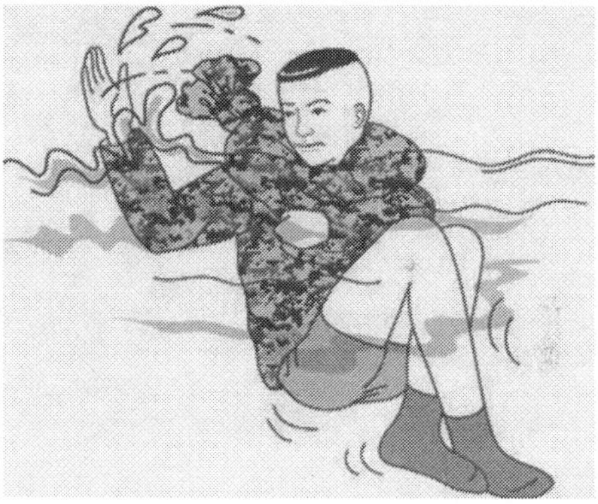

To replenish air in the trousers, you will use a technique known as the scooping method. With one hand on the open waistband, extend the trousers in front of you just below the surface of the water and scoop air bubbles with your free hand into the open waistband until the trousers have sufficient air. Repeat as necessary.

Splash Method

The splash method is an alternative to the sling method. As with the sling method, you must kick strongly to remain at the surface. To inflate trousers using the splash method, perform the following:

Take a deep breath, bend over, and remove your boots.

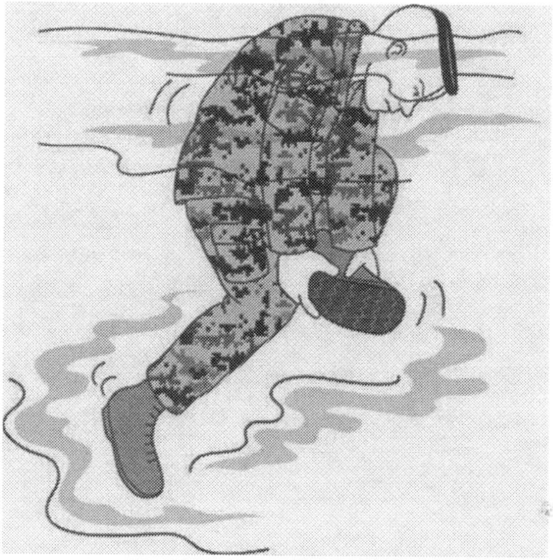

Note: Retain your boots. Tie the boot laces together and suspend the boots from your blouse or hang them around your neck so that they rest on your chest.

Remove your trousers. Button or zip the trouser fly closed. This allows you to control airflow.

Tie the bottoms of the trouser legs in a square knot. (The appendix illustrates various knots.)

Ensure that the front (fly) of the trousers faces you.

Hold the trousers at the water's surface out in front of you by the waistband with the fly up.

Grasp the waistband at the surface with one hand. Insert your free hand into the waistband, palm down.

Flutter your hand rapidly to create bubbles. This sends a mixture of water and air bubbles into the trousers. The water passes through the fabric. The air remains trapped in the legs.

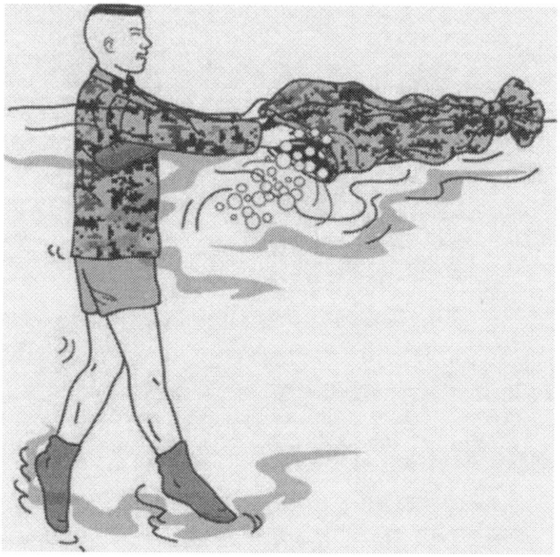

Hold and seal the waistband underwater.

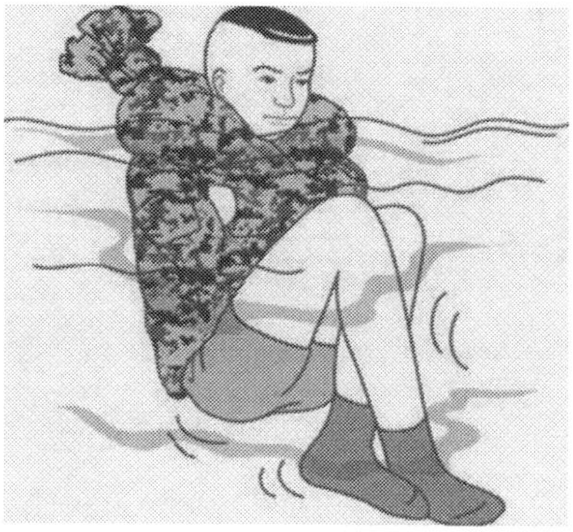

Slip the inflated legs over your head. Hold the waistband in toward your chest, the fly facing your body. To prevent air from escaping from the trousers, seal the waistband by either folding it or twisting it.

Lie back and relax, resting the back of your neck against the knot.

Splash water on the trousers periodically to prevent the material from drying. Dry material allows air to escape.

To replenish air in the trousers, you will use a technique known as the scooping method. With one hand on the open waistband, extend the trousers in front of you just below the surface of the water and scoop air bubbles with your free hand into the open waistband until the trousers have sufficient air. Repeat as necessary.

Blow Method

The blow method is an alternative to the sling method. Use the blow method if you are a weak swimmer. Take the following steps to inflate trousers using the blow method:

Take a deep breath, bend over, and remove your boots.

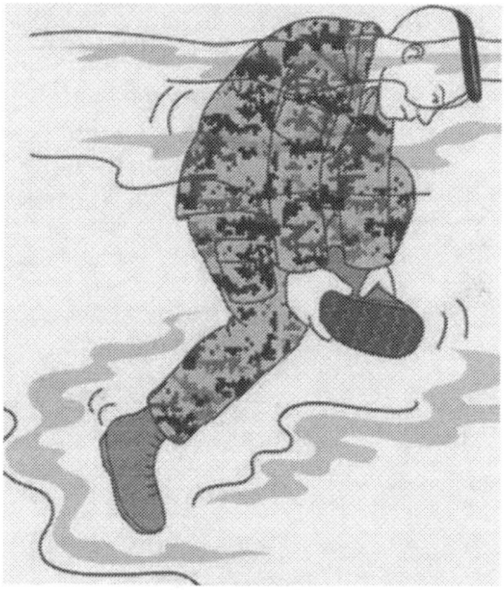

NOTE: Retain your boots. Tie the boot laces together and suspend the boots from your blouse or hang them around your neck so that they rest on your chest.

Remove your trousers. Button or zip the trouser fly closed. This allows you to control airflow.

Tie the bottoms of the trouser legs in a square knot. (The appendix illustrates various knots.)

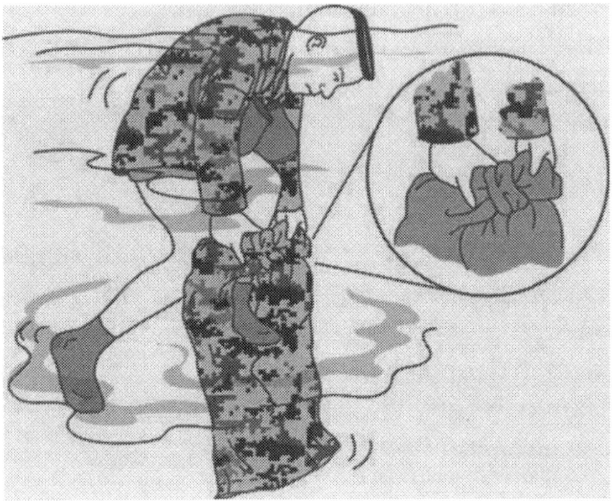

Ensure that the front (fly) of the trousers faces you.

Hold the trousers at the water's surface. Grasp both sides of the waistband and open with both hands.

Take a deep breath.

Drop 2 feet below the water's surface, pulling the waistband underwater.

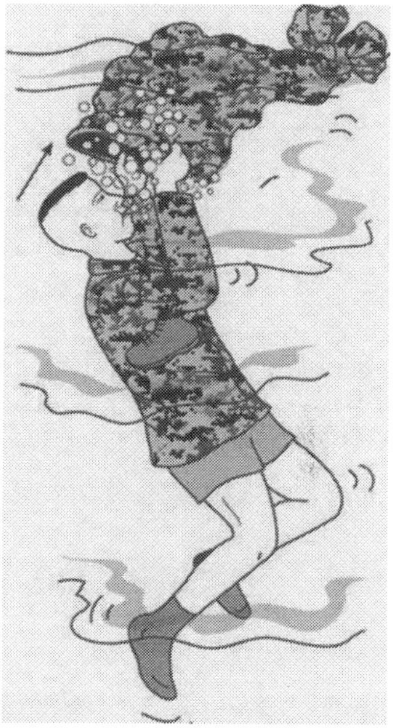

Hold the waistband open with both hands and blow air into the trousers.

To fill the trousers with air, surface while keeping the waistband underwater, breathe in again, drop below the water's surface, and blow air into the trousers. Repeat these steps until the trousers are filled sufficiently. Once trousers are filled—

Hold the waistband underwater. Twist and pinch it off.

Slip the inflated legs over your head. Hold the waistband in toward your chest, the fly facing your body. To prevent air from escaping from the trousers, seal the waistband by either folding it or twisting it.

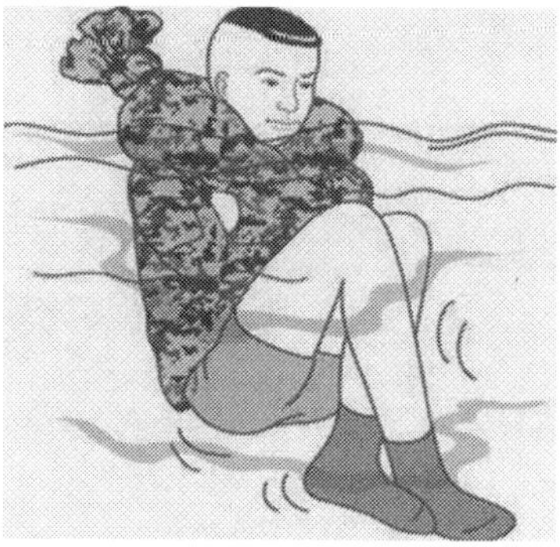

Lie back and relax, resting the back of your neck against the knot.

Splash water on the trousers periodically, to prevent the material from drying. Dry material allows air to escape.

To replenish air in the trousers, you will use a technique referred to as the scooping method. With one hand on the open waistband, extend the trousers in front of you just below the surface of the water and scoop air bubbles with your free hand into the open waist-band until the trousers have sufficient air. Repeat as necessary.

Avoiding Heat Loss in Cold Water

The rate of heat exchange in the water is about 25 times greater than it is in air of the same temperature. When you are immersed in cold water, hypothermia occurs rapidly due to the decreased insulating quality of wet clothing and as a result of water displacing the layer of still air that normally surrounds the body. You also lose about 50 percent of your body heat through your head; therefore, keep your head out of the water. Other areas of high heat loss are the neck, the armpits/sides, and the groin.

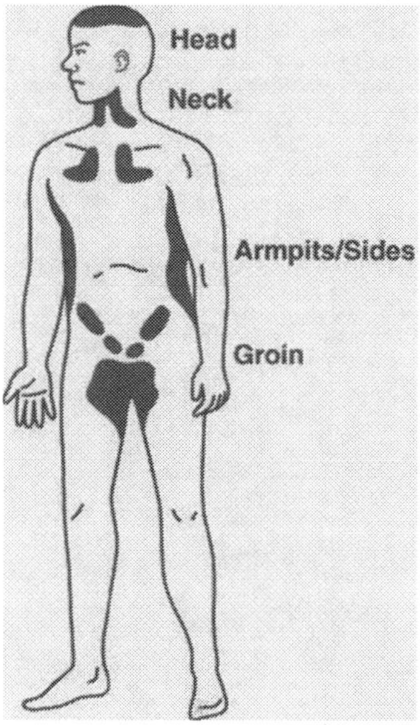

In cold water, DO NOT SWIM TO STAY WARM. Swimming, even with a slow and steady stroke, produces a lot of heat that is lost in the water. The heat loss can produce hypothermia that slows body functions and can result in serious injury or death. Remaining motionless conserves body heat three times longer than swimming. SWIM only if you have a flotation device and the shoreline is visible.

Individual Protection From the Cold

If you are equipped with a life preserver, assume the heat escape lessening posture (known as the HELP position) to slow heat loss and to protect major blood vessels near the body's surface. These areas lack insulating fat and are vulnerable to the chilling effects of cold water. To assume the HELP position—

Tuck your chin down tightly to cover your throat.

Draw your legs up in a fetal position to protect the groin.

Place your arms across your chest, tuck your hands into your armpits.

Wear some type of head covering (e.g., stowed cover, towel, handkerchief) to lessen heat loss through the scalp if head covering is available.

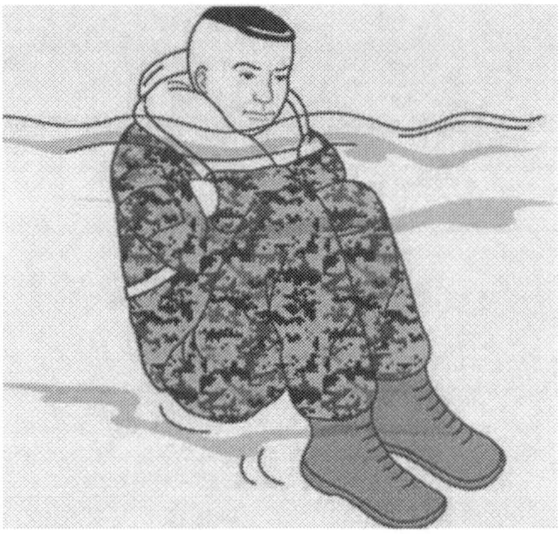

Group Protection From the Cold

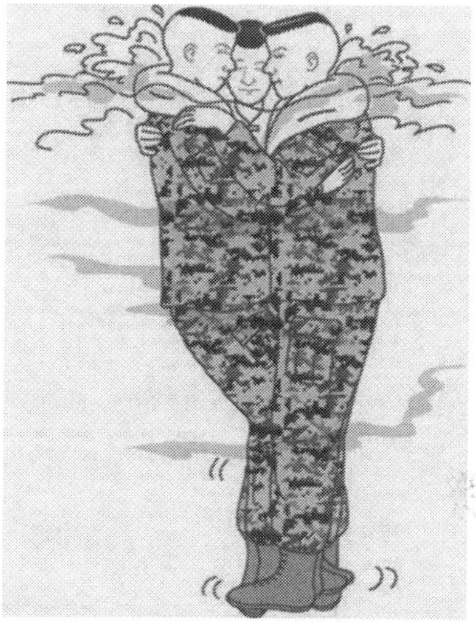

If three or more Marines are in the water and are equipped with life preservers, they should wedge tightly together and lock arms to form a circle known as a huddle position. This position protects vulnerable areas from heat loss.

A casualty who is suffering from the effects of the cold can be placed within the huddle to be surrounded by warmer water. If in the water for a prolonged period, it is recommended that everyone be rotated inside the huddle to maintain or re-warm each person's internal core temperature. If there are more than five Marines, they should make clusters of huddle positions.

Contact with other swimmers provides survival advantages—

Creates a larger target for search and rescue aircraft.

Provides additional warmth in cold water.

Improves morale.

Re-establishes the chain of command.

Reduces shock and panic.

Provides opportunities to administer first aid.

Supports exhausted Marines.

Drownproofing Methods

An object that floats on the surface has positive buoyancy. An object that floats a few feet beneath the surface has neutral buoyancy. An object that sinks has negative buoyancy. Most people have positive buoyancy and will float at the water's surface. To test your buoyancy—

Stand in water that is at your chest level or deeper.

Take a full breath.

Bend forward slowly.

Relax and wait.

If you have positive buoyancy, you will slowly rise to the surface. If you have neutral buoyancy, you will float a few feet beneath the surface. If you have negative buoyancy, you will sink. Regardless of whether you naturally float or sink, you can remain at the surface for extended periods without a life preserver if you exercise the appropriate drownproofing methods or survival strokes, which are based on your buoyancy.

Drownproofing methods consist of the T-method and the sweep. The breast stroke, side stroke, and elementary backstroke are the most common survival strokes. With any drownproofing method or survival stroke, remember the acronym **SAFE**:

SLOW EASY MOVEMENTS
Move slowly to conserve energy and minimize heat loss.

APPLY NATURAL BUOYANCY
Use natural buoyancy to support the body.

FULL LUNG INFLATION
Fill the lungs with each breath. Do not hold air in the cheeks.

EXTREME RELAXATION
Tight muscles are denser than relaxed ones and do not float as well.

Crawl Stroke

The crawl stroke, sometimes called the front crawl or free style, is the fastest stroke. To execute the crawl stroke—

Body position
Lie horizontal, on your stomach, in the water.

Look forward and downward at a 45 degree angle with the waterline between your eyebrows and hairline. Your head position is important as it assists in cutting a path through the

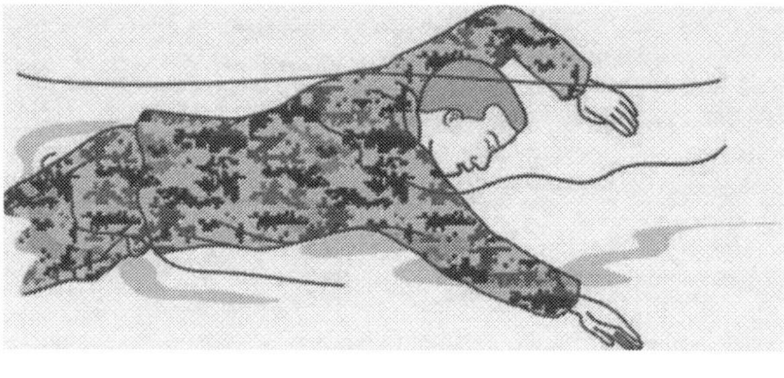

water. If your head is too high in the water, your lower body sinks significantly, making your stroke less efficient. If your head is too low in the water, water washes over your shoulders and neck, causing unnecessary drag.

Arm action

Arm action occurs in three phases: catch, propulsion, and recovery.

Fully extend one arm forward of your body, this positions your hand to catch the water in preparation for the propulsion phase.

To catch the water, bend your wrist (with your palm pointing outboard) and make an "S" shape (or inverted "S" shape) with your hand, ensuring that your hand does not cross the center of your body. Your left hand makes an "S" shape and your right hand makes an inverted "S" shape with your hand finishing at shoulder level.

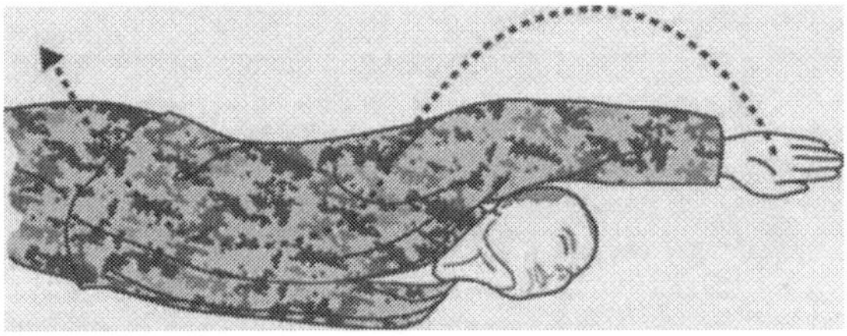

Push with your hands in a rearward fashion toward your feet until your arm is fully extended along your side, keeping your hands close to your body.

To begin the recovery phase, bend your arm at the elbow and raise your hand out of the water. Your hand breaks the surface of the water and maintains a height of 2 to 3 inches above the water's surface.

With your hand and arm moving in a forward manner, bring your hand past your head until your arm is about three-quarters of the way extended. At this point, turn your hand with your palm outboard so as to allow your thumb and fore-finger to enter the water first.

Once your hand enters the water, continue to push your arm forward until it is fully extended.

You are now prepared to catch the water again. These steps are performed in an alternating pattern: when one arm is catching and propelling, the other arm is recovering.

Leg action
Use a flutter kick to create the leg action for the crawl stroke. This kick is used for both propulsion and keeping the lower body horizontal with the water's surface. The flutter kick is an alternating leg action: one leg is kicking in a downward motion while the other leg is recovering to the surface to prepare for the next kick. The size of the kick ranges from 12 to 15 inches and depends on your height.

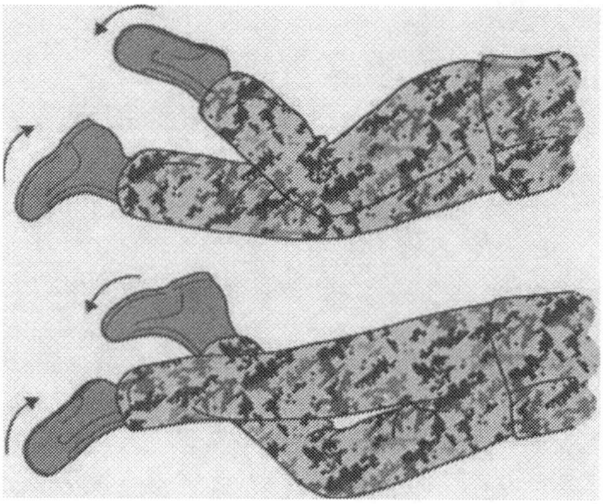

Maintain your legs in a semi-rigid manner.

Generate power for the kick from the hips.

Keep your feet loose at the ankles, they trail behind your legs and act as "flippers."

Execute the propulsion phase of this kick with a downward thrust of your leg.

Execute the recovery phase by pushing your leg back to the surface. This phase is complete when your foot reaches the surface.

Breathing
Breathe during the recovery phase of the arm action.

Roll your body and rotate your head to the side of your body where the arm recovery is occurring, this rolls the water away from your mouth. Keep your chin pushed back toward your shoulder.

Exhale while your face is still submerged and inhale when your face breaks the surface.

Breathe either bilaterally or rhythmically. To breathe rhythmically, breathe on the same side of your body every time your arm cycle occurs. Bilateral breathing requires you to breathe every one and a half arm cycles. To breathe bilaterally, breathe when your right arm is recovering, your face goes back in the water and the next time you breathe is when your left arm is starting its recovery three arm strokes later.

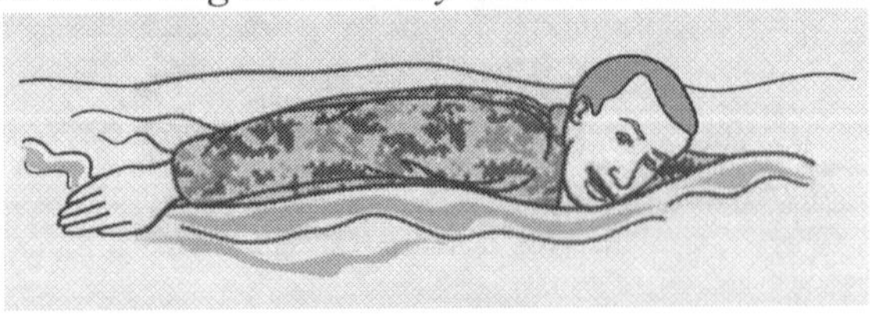

Bilateral breathing is the preferred method for breathing; it prevents the chance of hyperventilation and allows your body to maintain a lateral position in the direction you are swimming.

Coordination
This stroke uses constant arm and leg action.

T-Method

The T-method is a basic drownproofing method. This is the best survival technique if you have negative buoyancy. To execute the T-method—

With your face out of the water, take a deep breath and submerge your face in the water while holding your breath.

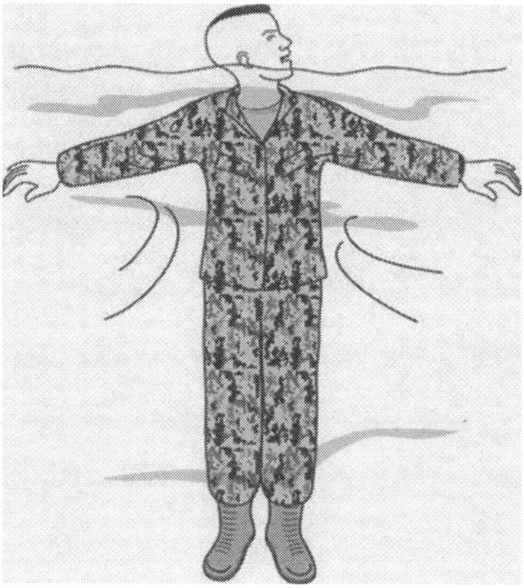

Float with your body in a horizontal position, arms extended from your side, and legs extended and joined.

Move your hands up to your armpits by tracing an imaginary line along your ribs.

Extend your arms outward (horizontal) to your sides, your body position resembles the letter "T."

Step out and forward with one leg and point your other leg to the rear, your knees should be slightly bent. Simultaneously, bring your arms down to your sides. Then bring your legs back together. You complete the steps by exhaling most of your air and preparing to surface your face to obtain another breath of air. Hold your head out of the water. Tilt your head back slightly. Breathe normally.

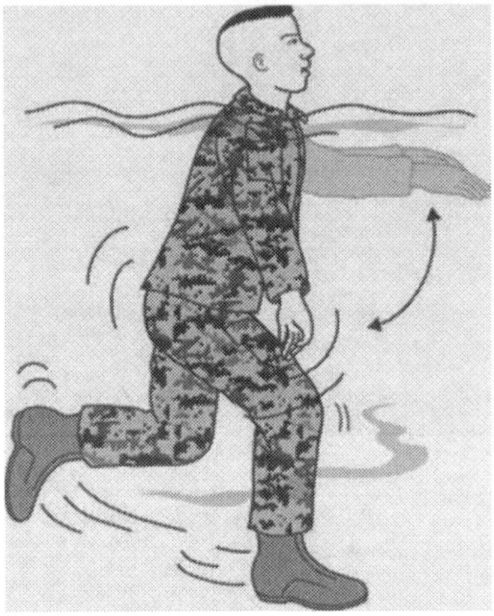

Once your breath is complete, move your hands up and down directly in front of your body. Do this two or three times to slow your descent into the water.

> *Note:* To avoid hyperventilating, hold your breath below the surface of the water for no more than 10 seconds.

The Sweep

The sweep works well if you have slight to excellent positive buoyancy. To execute the sweep—

Float face down in the water: bend 45 degrees at the waist, arms and legs dangling, head hanging down, relax all muscles.

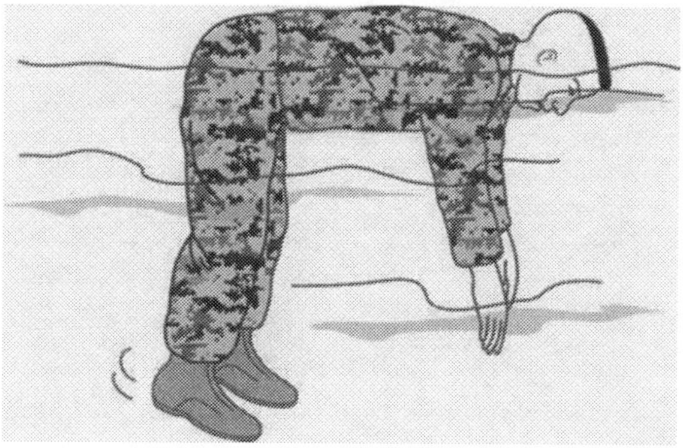

Spread your feet slowly to prepare for a single kick (one leg is forward and one leg is rearward).

Cross your arms in front of your chest, palms outboard with the back of each hand touching the opposite ear.

Exhale prior to raising your head for a breath.

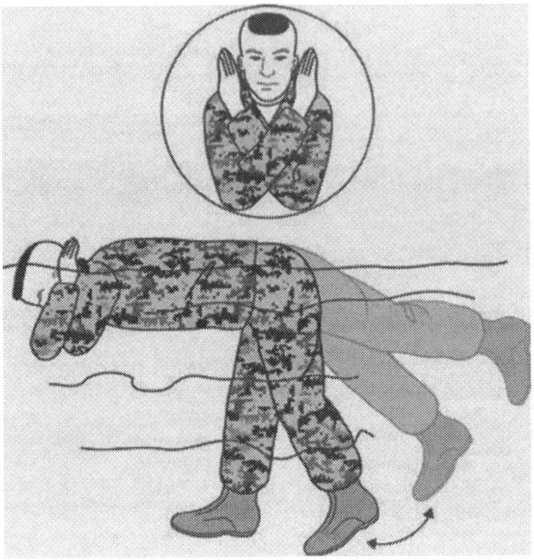

Bring your legs together and sweep your arms down and out until the arms are fully extended out to the sides. This raises your face above the water and allows you to catch a breath of air.

With your air supply replenished, return your face to the water and relax while sweeping your arms in a downward motion in front of your body to prevent/slow your descent. Do not hold your breath for more than 10 seconds. More than 10 seconds enhances your chances of shallow water black out and subsequent drowning.

Breast Stroke

Use this stroke to swim underwater, through oil or debris, and in rough seas. If you are a good swimmer and not wearing combat gear, the breast stroke is the best stroke for long-range swimming because it provides good visibility and allows you to conserve your energy and maintain a reasonable speed.

Body position
Lie prone in the water. Swim with your trunk and legs projecting back and down at an angle of 20 to 30 degrees.

Extend arms out in front (hands together [side by side]), and extend legs behind (toes pointed) to prevent drag.

Face downward, looking forward at a 45 degree angle to break the water and to prevent water from washing into the collar area causing drag.

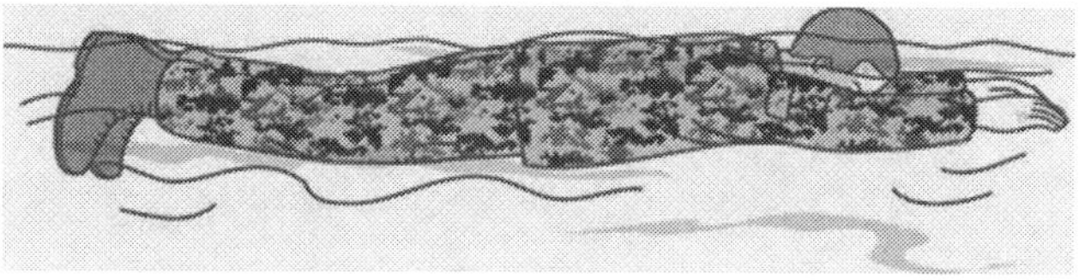

This is known as a glide.

Arm action

Turn your palms outward and bend your arms slightly.

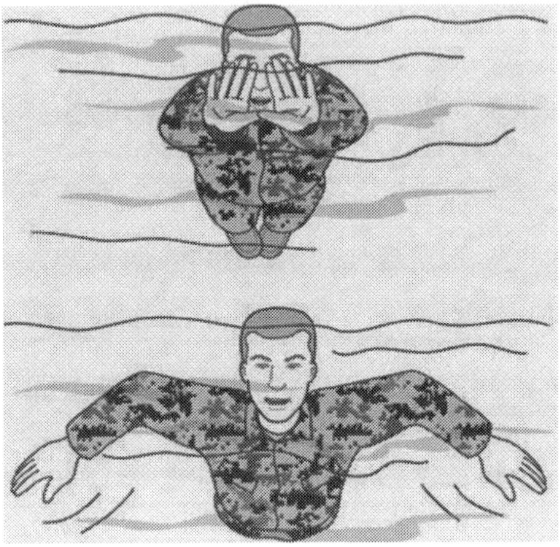

Sweep your arms sideward and slightly downward until your hands are opposite and slightly below your shoulders.

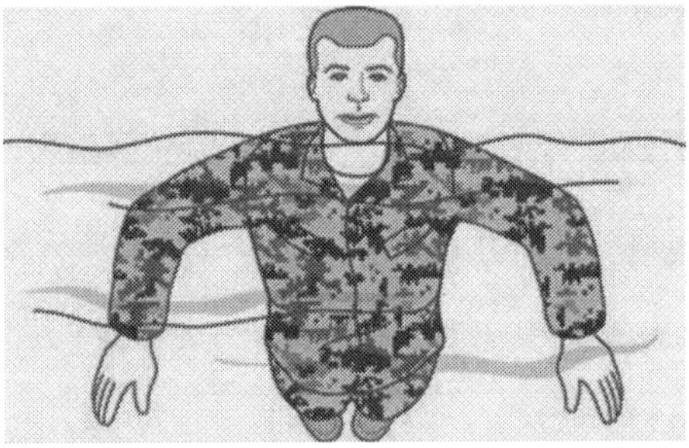

Rotate your head up, breathe once your mouth breaks the surface.

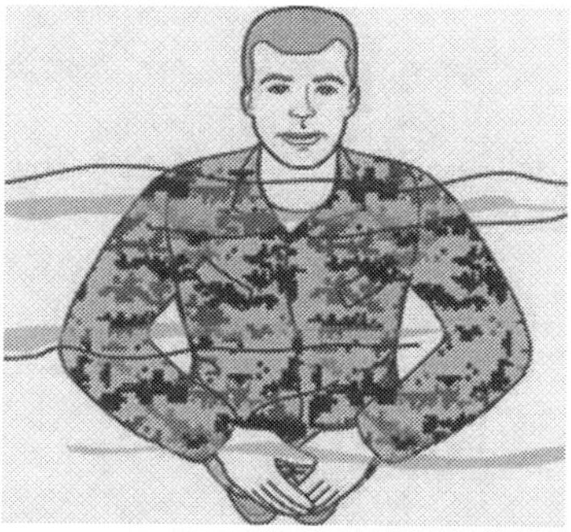

Bring your hands and arms up along your chest and thrust them forward until they are extended and ready to execute the next arm pull.

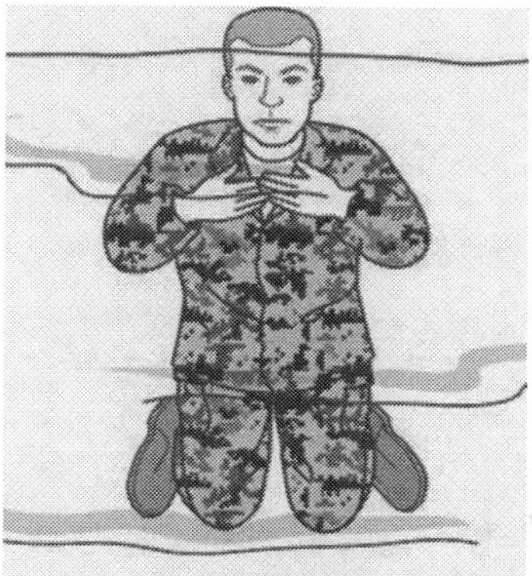

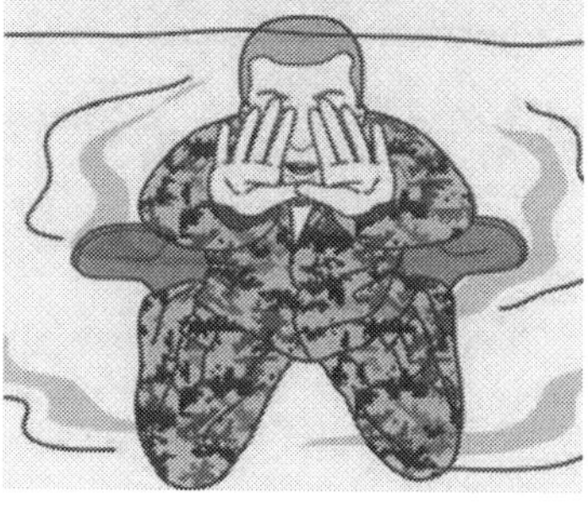

As the arms start their recovery into the glide, the head should rotate forward, resubmerging the face.

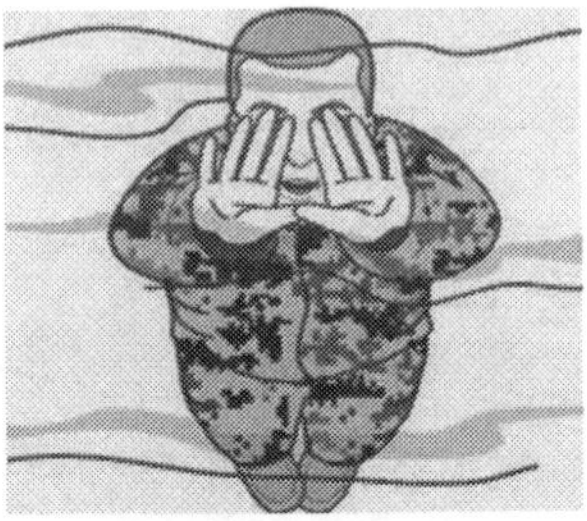

Leg action

Draw your heels toward your buttocks, establish a 45 degree bend in the knees.

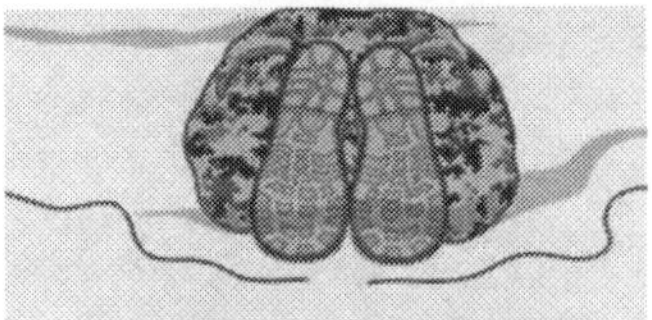

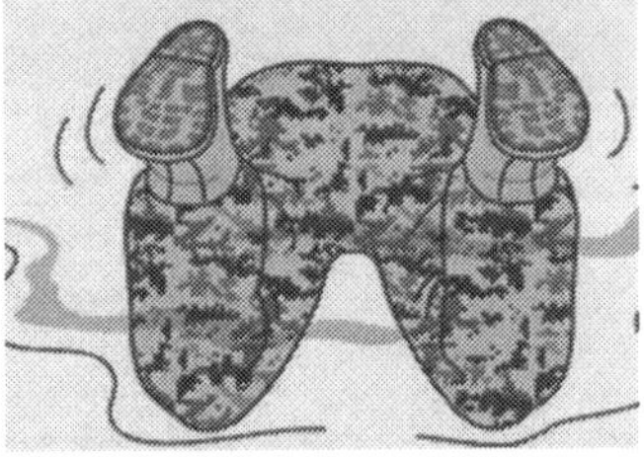

Thrust your legs outward and rearward, then squeeze them together. The whipping action of the feet aids forward propulsion.

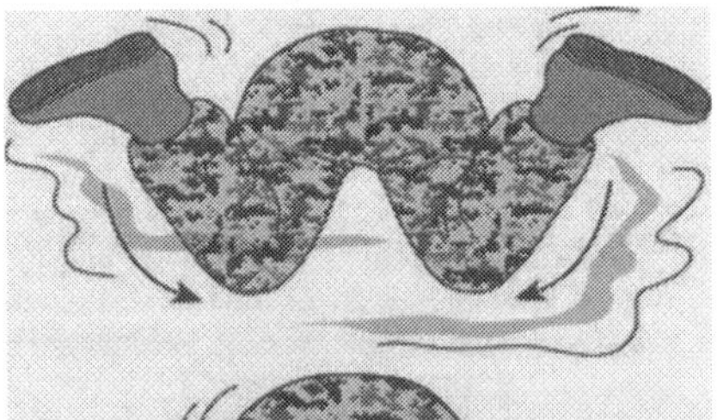

This is known as the breast stroke kick.

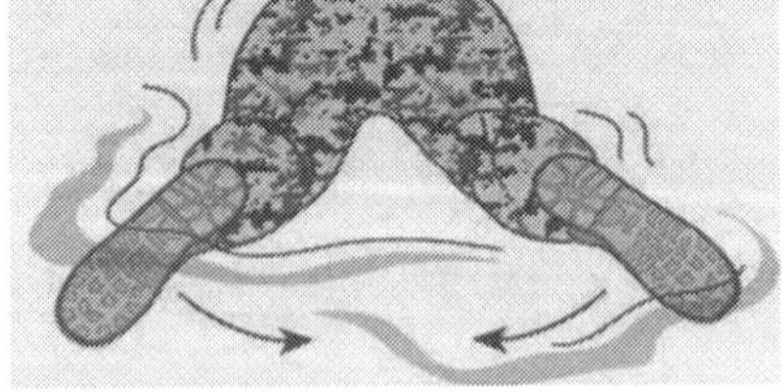

Breathing
Inhale during the arm pull and exhale through your mouth and nose during the finish of the breast stroke kick and glide.

Coordination
The stroke movement is in three counts:

Begin your arm pull. Near the finish of the pull, flex your knees and bring your heels toward your buttocks. The arm pull counteracts the resistance created by the knees.

As the arm pull is completed, thrust your hands forward, kick your legs outward and rearward, and squeeze them together.

Glide through the water for approximately 1 to 3 seconds or until your forward momentum decreases, then begin the next stroke.

Side Stroke

The side stroke is a survival stroke because you use both arms for buoyancy, with each arm creating a slight propulsion. The majority of your body's propulsion comes from your kick. To execute the side stroke—

Body position
Lie on your side with your lead (bottom) arm extended beyond (with a slight bend in your elbow) your head and in line with your body. Palm is down and your hand is submerged 6 to 8 inches.

Extend your trail (top) arm down the length of your body over your thigh.

Keep your legs straight and together, toes pointed rearward.

Keep your face out of the water, this allows for free breathing.

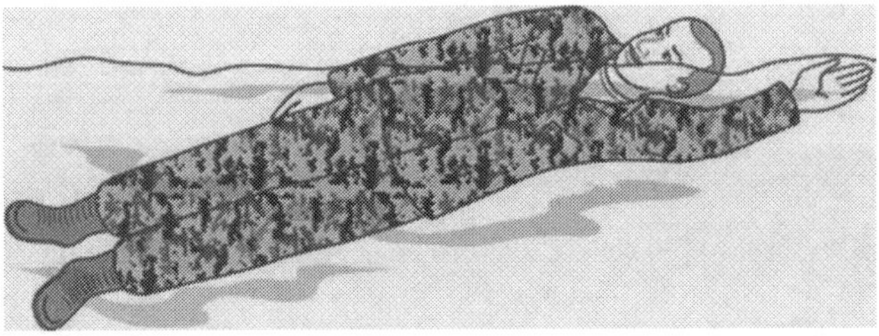

This is known as the glide.

Arm action

With your lead arm, pull your arm downward, while flexing at the elbow, until it is straight down from your shoulder.

Rotate your shoulder and pull your elbow into your side. This should put your lead hand at shoulder level. At the same time, turn your palm toward your face and thrust forward to your original, extended position.

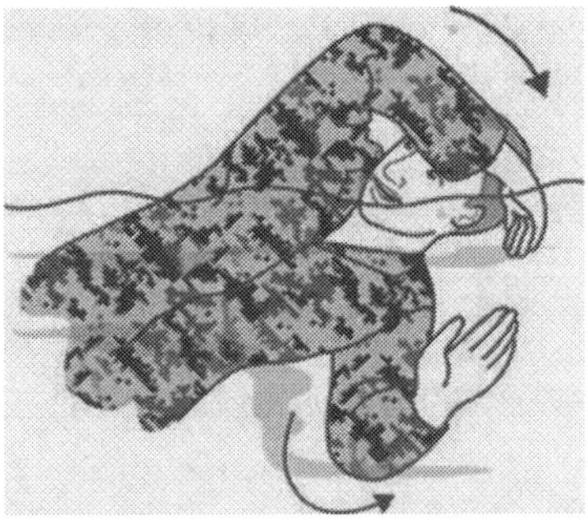

Draw your right hand upward in front of your chest to shoulder level. Rotate your palm toward your feet, then push it downward in front of your body toward your feet to catch the water.

Push your trail hand backward to its original position on top of your thigh. (Your trail hand starts forward and meets your lead hand at your chest/shoulder.)

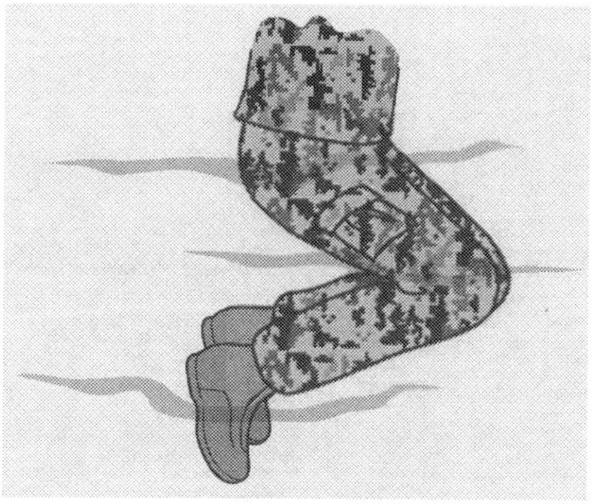

Leg action
To perform the scissor kick, the top leg always goes forward and the bottom leg always goes rearward. From the extended position, draw—or recover—your feet toward your buttocks until your legs are bent at a 45 degree angle at the knees and the hips are flexed at a 45 degree angle with the thighs.

Once the legs have completed their recovery and while maintaining a 45 degree bend in the knees, extend the legs fully into a "V" shape in order to catch the water for the propulsion phase.

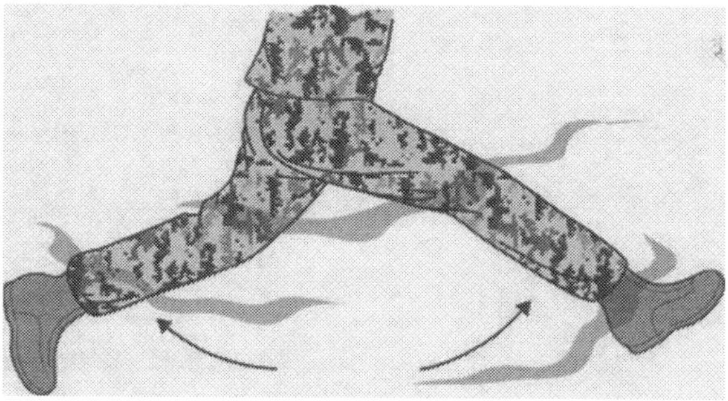

Once the legs are separated and extended forward and rearward to the "V" position, sweep the legs together until the feet are together.

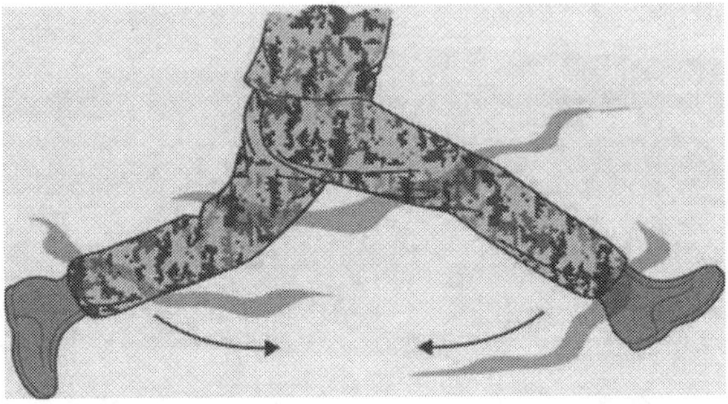

You are now in the glide position.

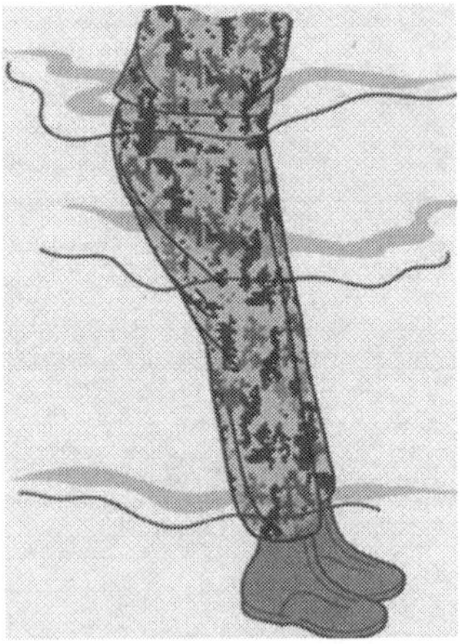

Breathing

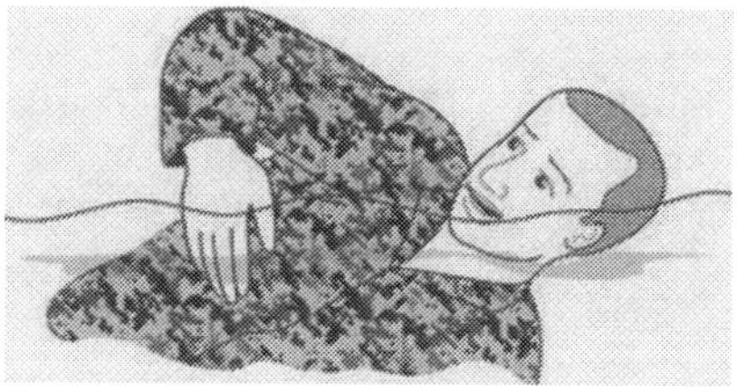

As long as your face remains clear of the water, it is a free breathing stroke. However, it is recommended that you exhale then quickly inhale when the legs are sweeping back together in the scissor kick. This is when the body reaches its highest point in the water, thus clearing the face completely from the surface of the water making it the optimum time to breathe.

Coordination

Begin the stroke with the downward pull of your lead arm. At the same time, bring your trail arm upward and draw your knees up to begin the kick. Let the thrust of the lead arm, push of your trail arm, and the kick of your legs coincide in order to finish the glide position. Glide through the water for approximately 1 to 3 seconds or until your forward momentum decreases, then begin the next stroke.

Elementary Backstroke

The elementary backstroke is also an excellent survival stroke. It relieves the muscles that you use for other strokes, and it is the recommended stroke for weak swimmers or nonswimmers. To execute the elementary backstroke—

Body position
Start on your back.

Face up, chest up, and hips up, keeping an arch in your lower back with arms pressed to your sides and your legs extended and joined to prevent drag.

Arm action
Trace your hands up your sides to an area near your armpits then extend your arms out to the sides to form the letter "T" (palms facing feet), locking out the elbows.

> NOTE: Don't raise your arms above your head. This creates drag, changes your body position, and submerges the head.

Slap your palms to your thighs using a strong sweeping motion.

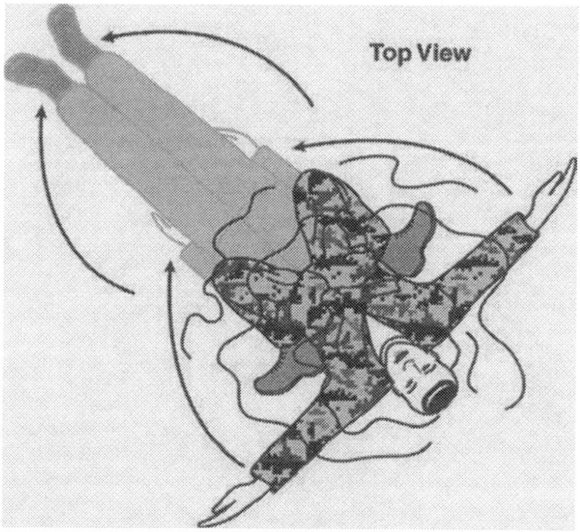

Leg action

Bend both legs at the knee (90 degree angle) slightly separating your knees and drawing your heels downward to a point under and outside your knees. The knees are spread as wide as the hips or slightly wider depending on the body type of the swimmer.

Circle around in a whipping action, ending with legs in a glide position.

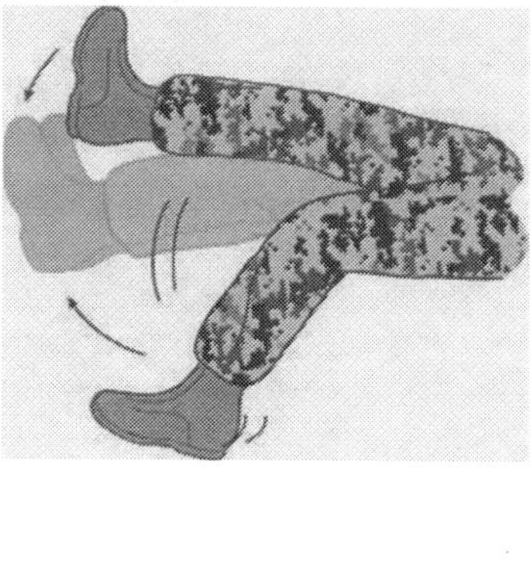

Breathing

Breathe anytime during this stroke. However, it is recommended that you exhale then quickly inhale when your arms are sweeping back toward your sides and while your legs are sweeping back together. This is when the body reaches its highest point in the water, thus clearing the face completely from the surface of the water and making it the optimum time to breathe.

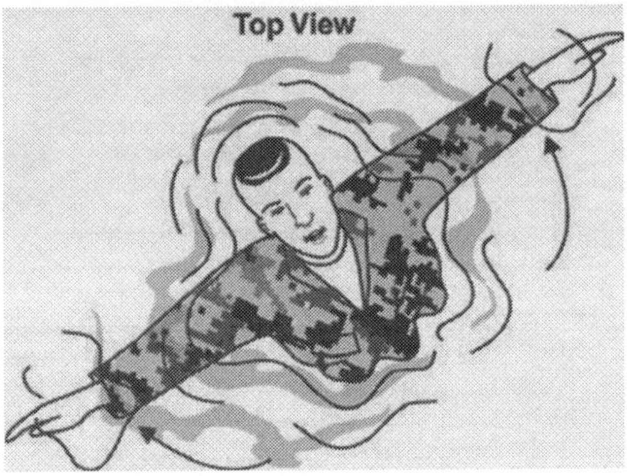

Coordination

The stroke movement occurs in three counts (recovery, catch, power).

Begin the arm pull (recovery).

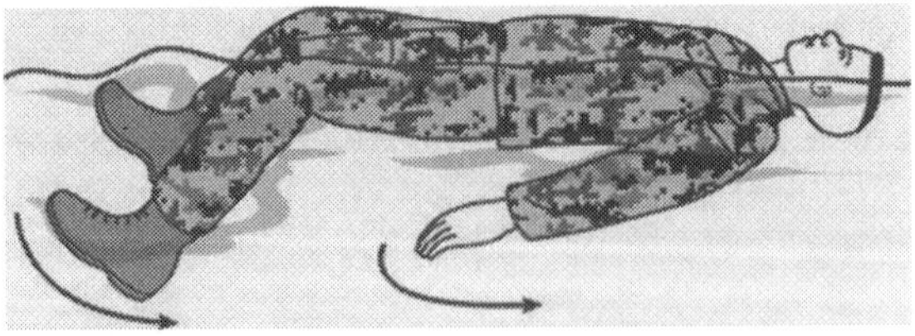

Near the finish of the pull, flex your knees to a 90 degree angle. The arm pull counteracts the resistance created by the knees.

Kick out your legs, and squeeze them together as the arm pull is completed (catch, power).

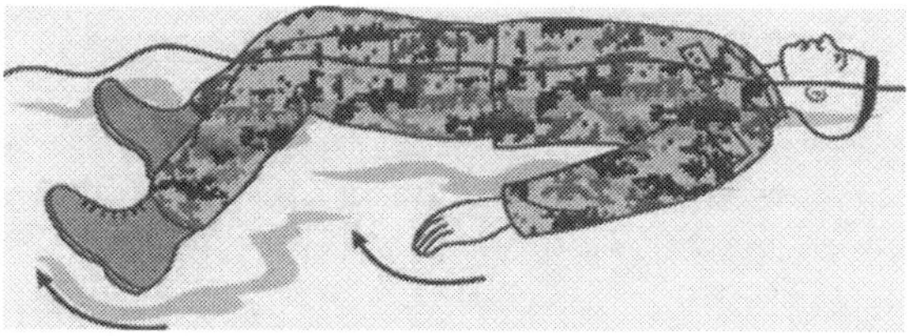

Glide through the water for 1 to 3 seconds or until your forward momentum decreases, then begin the next stroke as your momentum slows.

Chapter 2
Water Rescues

A drowning victim can panic and react with unexpected violence and can seize and inadvertently drown a rescuer. Therefore, if possible, a water rescue should be executed from a distance. Reaching, wading, or throwing methods are used in a distance rescue.

If a victim is too far away to use these methods, a swimming rescue may be necessary. Remove all combat gear, when possible, before entering the water. Swim within 2 to 6 yards of the victim to maintain a margin of safety, this allows you to reassess the situation and reassure the victim. If the victim is unconscious, use the wrist tow method or cross-chest carry method to pull the person to safety. If the victim is struggling, use a rear approach and then execute either a single armpit level off or a double armpit level off before towing the victim to safety.

If the victim does begin to overpower you, there are techniques that allow you to defend yourself without having to abandon the rescue. These techniques include the block, the wrist-grip escape, the front head-hold escape, and the rear head-hold escape. These techniques allow you to separate yourself from the victim, reassess the situation, and then attempt the rescue again.

> *NOTE:* In this chapter's illustrations, the rescuer is shown without a helmet.

Reaching Rescue Techniques

Reach

From a safe position at the water's edge, reach out to the victim. Talk constantly to calm the victim. Retain partial contact with land or some solid support structure (e.g., pier, bridge). If the victim is close but still beyond reach, extend an object (e.g., stick; pack; rifle with magazine removed, chamber empty, and muzzle pointing toward victim) that the victim can grasp. Pull the victim slowly to safety. Once the victim is close to shore, remove the victim from the water.

Note: You can also extend a foot to the victim if you can retain a secure grip on a solid support structure.

Reach From a Deck

Reach from a deck rescue can be executed by either a swimmer or a nonswimmer, and it can be used on an active or a passive victim. Be sure to reassure the victim during the rescue. To execute reach from a deck—

Lie prone on the deck with your body firmly anchored. To anchor your body, lie flat, spread legs apart, and extend one arm behind you with your palm down and on the deck.

Keep as much of your weight on the deck as possible and extend your free hand to the victim.

Grasp the victim's wrist from above, your thumb and index finger are facing you. The victim should never be allowed to grab you and put your life at risk. Therefore, never reach across the victim to grab his wrist, always grab the wrist that is the closest to you.

Keep your arm straight and locked out and pull the victim to the side of the deck. The victim should never be allowed to grab you and put your life at risk. Therefore, never pull the victim into you, always pull the victim into the side.

Arm Extension

An arm extension rescue can be used if you cannot reach the victim using the reach from a deck rescue technique and you must enter the water. An arm extension rescue technique is used for a victim who is either active or passive. Once you determine that a reach from a deck rescue technique is not viable—

Reassure the victim and quickly ease into the water while holding onto the deck with one hand.

Grasp the victim's wrist from above, your thumb and index finger are facing you. The victim should never be allowed to grab you and put your life at risk. Therefore, never reach across the victim to grab his wrist, always grab the wrist that is the closest to you.

Keep your arm straight and locked out and pull the victim to your side. The victim should never be allowed to grab you and put your life at risk. Therefore, never pull the victim into you, always pull the victim into the side.

Leg Extension

If the victim is beyond the reach of your arm, ensure that you have a firm grip on the deck, extend a leg to the victim and allow him to grab it, slowly bring the victim in closer until you can grab the victim's wrist that is holding onto your leg. Once you have a firm grasp on the victim's wrist, use the steps in the arm extension to pull the victim to safety. The leg extension rescue can only be used to rescue an active victim because the victim must be able to grab the rescuer's leg.

Wading Assist

Do not wade into water that is deeper than your chest. Talk constantly to calm the victim. If possible, do not touch the victim directly. Extend an object (e.g., stick; pack; rifle with magazine removed, chamber empty, and muzzle pointing toward victim) that the victim can grasp. Once the victim grasps the object, pull the victim slowly to safety.

Throw

If the victim is not within reach, use an expedient line to throw a lifesaving device to the victim. A lifesaving device can be any weighted item that floats (e.g., a canteen that is one fourth full of water). The lifesaving device must be secured to the end of the rope so that the rope will feed out from its coil when tossed. Talk constantly to calm the victim. Once the victim grasps the line or the lifesaving device, pull the victim toward you at a steady pace that keeps the victim's head above the water's surface. DO NOT pull so strongly that you break the victim's grip on the line. To prepare and use an expedient line and lifesaving device—

Tie a bowline at one end of the rope. (The appendix illustrates various types of knots.)

Unfasten the lid of a canteen.

Place the bowline around the neck of the canteen.

Refasten the lid so the canteen hangs from the bowline loop.

Place one end of the rope under the ball of your forward foot to secure it, either tie a knot at the end of the rope or tie an object at the end of the rope to create a block. Stand with your weight on the end of the rope. DO NOT tie the rope around your ankle.

Coil 20 to 30 yards of the rope, and hold it in a nonthrowing hand.

Place the canteen in your throwing hand.

Use an underhand throw to pitch the canteen and rope a short distance over the victim's head. Keep your nonthrowing hand open so the coil can unfold freely. The rope should trail across the victim's outstretched arms.

Retrieve the rope if the throw is inaccurate or the victim fails to grasp it. Recoil the rope as it is retrieved.

Divide the coil and throw again.

Lifesaving Approaches

Properly approaching the victim is as important as any other aspect of a rescue. Knowing the appropriate and safest method needed to approach a victim is critical in ensuring your safety and the victim's survival. Determining the victim's physical state (distressed swimmer, victim that is active and drowning, or victim that is passive and drowning) is crucial and will determine which type of approach you will execute.

Front Surface Approach

The front surface approach is typically performed when the victim is passive. You must remember that approaching the victim from the front is dangerous because a distressed swimmer or victim that is active can lunge toward you.

Rear Approach

If the victim is active, you must remember that his response can change rapidly. Therefore, approaching him from the front could be extremely dangerous. The victim could easily grab you and take you under the water during a state of panic. Therefore, use a rear approach when possible. When approaching a victim from the rear, reassure the victim until you get to within 2 to 6 yards of the victim; at that point, stop talking to the victim in order to maintain an element of surprise.

Approach Strokes

Before approaching a victim, you must evaluate the victim:

- Is the victim a distressed swimmer?
- Is the victim active and drowning?
- Is the victim passive and drowning?

Select your approach stroke based on your evaluation. Remember, the safety of the rescuer is paramount. Do not endanger yourself in an attempt to reach a drowning victim.

Crawl Stroke Approach Stroke

The crawl stroke (free style) is the fastest approach stroke. This stroke is used when the victim is passive and/or unconscious in the water. That is, the victim is face down, submerged or near the surface of the water, is not breathing, or is not moving. If any of these conditions exist, there is an obvious need to reach the victim as quickly as possible.

Execution of the crawl stroke approach stroke is the same as the crawl stroke (see pages 1-40 through 1-44), except that your head remains above the water's surface to allow for free breathing and maintaining eye contact with the victim.

Maintaining eye contact is critical. If the victim becomes submerged, you will have a better chance of locating him if you know his last location. Your distance from the victim will determine if your head remains out of the water the entire time you are executing the stroke. If you are 54 yards or less from the victim, you should maintain eye contact by keeping your head raised until you reach the victim. If you must swim more than 54 yards to reach the victim, make eye contact with the victim, place your face back in the water for four or five strokes, then raise your head again to regain eye contact with the victim (repeat these steps until you reach the victim).

Body position
The position of your body is horizontal to diagonal and influenced by the position of the head. This allows you to maintain eye contact with the victim.

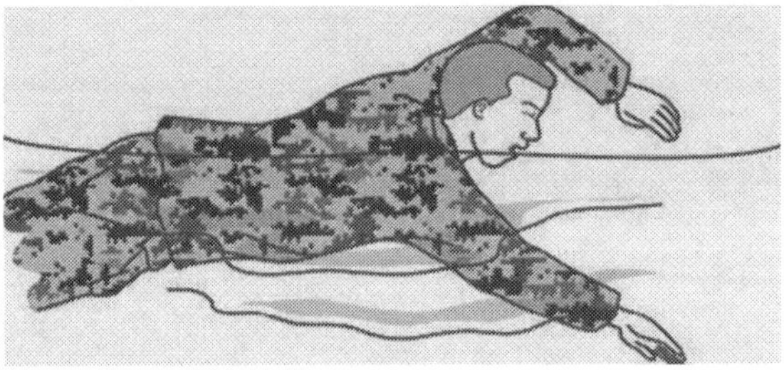

Arm action
While one arm is forward of your head to catch the water for the propulsion phase, extend your other arm alongside your body in what is known as the recovery phase. The recovery phase ends when your arm is extended forward of your head to begin the propulsion phase. In the propulsion phase, your arms are approximately shoulder-width apart and the pulling action is slightly wider and deeper to compensate for your raised head.

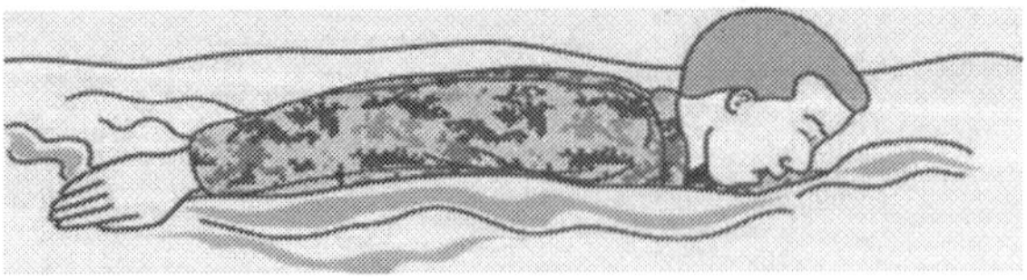

Leg action
Kick one leg in a downward motion while your other leg is recovering to the surface to prepare for the next kick (this is known as a flutter kick). This kick is used for propulsion and maintaining the lower body as horizontal to the water's surface as possible. When your head is raised, your knees are slightly bent in order to keep your feet near the surface.

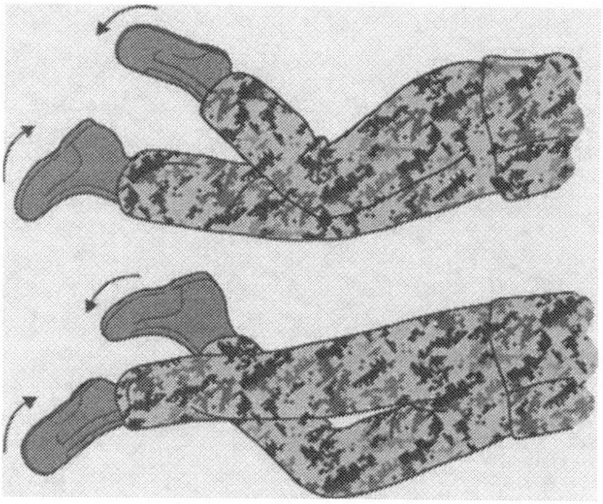

Breathing
If your face is above the water, breathe as needed (free breathing).

If traveling in excess of 54 yards to reach the victim, rotate your head to the right or left during the recovery phase and inhale when your face is out of the water or wait until your head is out of the water and facing forward while regaining sight of the victim.

Coordination
This stroke uses constant arm and leg action.

Breast Stroke Approach Stroke

The breast stroke approach stroke is used when a swimmer is distressed or a victim is active and drowning. This stroke is also used when the victim has a suspected spinal injury because it minimizes the wake created around the victim, minimizes the movement of the victim's head and neck, and helps prevent further injury.

If a victim is at a distance greater than 54 yards and if the swimmer is in distress or the victim is active and drowning in open water (ocean, lake), you should use the crawl stroke approach stroke to rapidly approach the victim, stopping approximately 11 yards from the victim in order to assess the situation.

Execution of the breast stroke approach stroke is the same as the breast stroke with a few minor deviations.

Body position
The position of your body is horizontal to diagonal and influenced by the position of the head. This allows you to maintain eye contact with the victim. It also allows you to communicate with the victim in order reassure him as you approach.

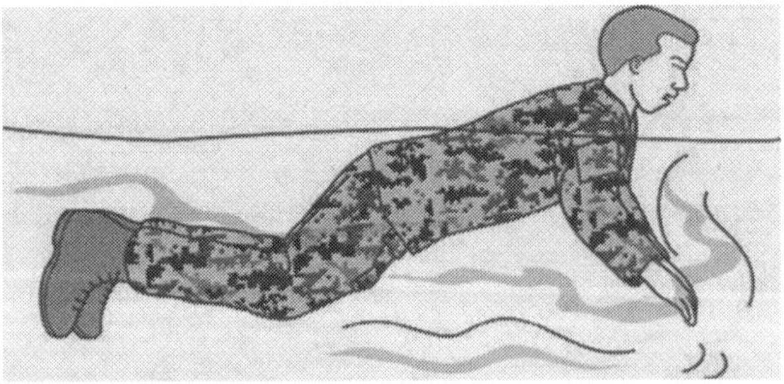

Arm action

Execute the arm action for the breast stroke, except that your arm pull is wider and deeper, which allows your head to remain above the water's surface.

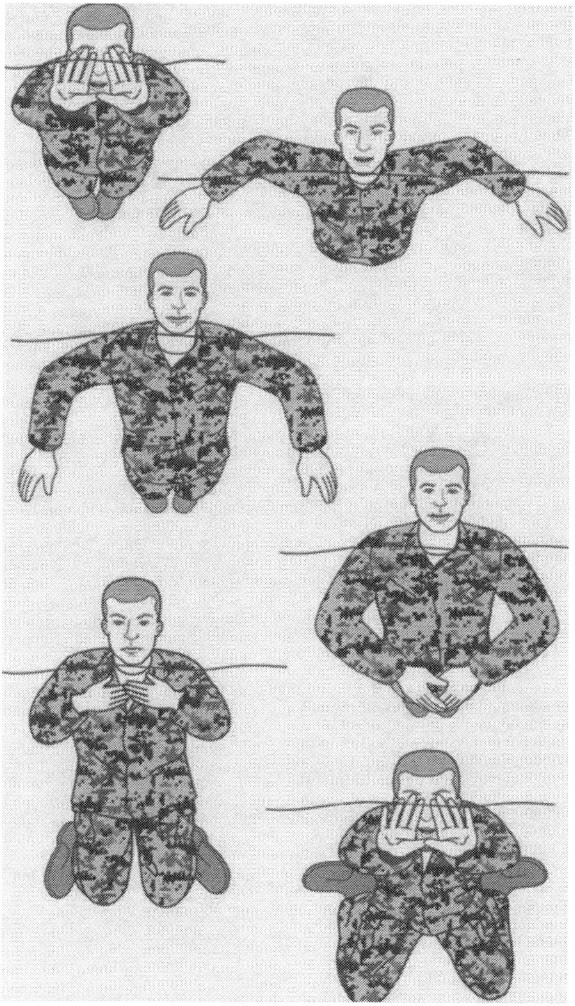

Leg action
Execute the leg action for the breast stroke, except that your legs open wider during the propulsion phase of the stroke.

Breathing
Keep your face above the water and breathe as needed (free breathing).

Coordination
Coordination is the same as the breast stroke, except that there is no glide period. You constantly stroke until you reach the victim.

Level Offs

A properly performed level off positions the victim's face up and horizontal to the water's surface. There are three types of level offs: front surface approach, single armpit level off, and the double armpit level off. The single armpit level off and the double armpit level off are done by performing a rear approach, which is the preferred approach when a victim is active.

Front Surface Approach

The front surface approach is performed when the victim is passive. Approaching a victim from the front may place you in danger because a distressed swimmer or a victim who is active may lunge toward you. To execute a front surface approach—

Stop 2 to 4 yards from the victim to reassess the situation and reassure the victim.

Determine which arm to use in order to rotate the victim into a face-up position (your right hand on the victim's right wrist or your left hand on the victim's left wrist).

Move into position and prepare to grasp the victim.

Turn sideways and move toward the front of victim.

Once in position, reach forward and grab the victim's wrist. Your thumb is on the inside of the victim's wrist, as if you were checking his pulse with your thumb. Your remaining fingers wrap around the victim's wrist.

Lean back immediately and execute a powerful scissor kick or inverted scissor kick and perform short, vigorous pulls with your free arm.

As the victim begins to move forward in the water, kick, pull, and twist outboard on the victim's wrist. The momentum created from the kick and the pulling and twisting action of the victim's wrist will rotate the victim into a face-up and horizontal position in the water.

Extend and lock out your towing arm down the length of your body and execute a wrist tow to move the victim to safety.

Single Armpit Level Off

Approach the victim.

Stop 2 to 6 yards from the victim to reassess the situation, reassure the victim, and maintain your margin of safety.

Approach the victim slowly and grab his armpit with your hand (your right hand to victim's right armpit or your left hand to victim's left armpit). Position yourself sideways to the victim and place your elbow in the center of his back.

Pull with your hand that is in the victim's armpit while pushing with your elbow that is in the victim's back to place the victim horizontal on the water's surface (face up). To assist in placing the victim horizontally, use your free arm to execute short, vigorous pulls and use your legs to execute a scissor kick or inverted scissor kick.

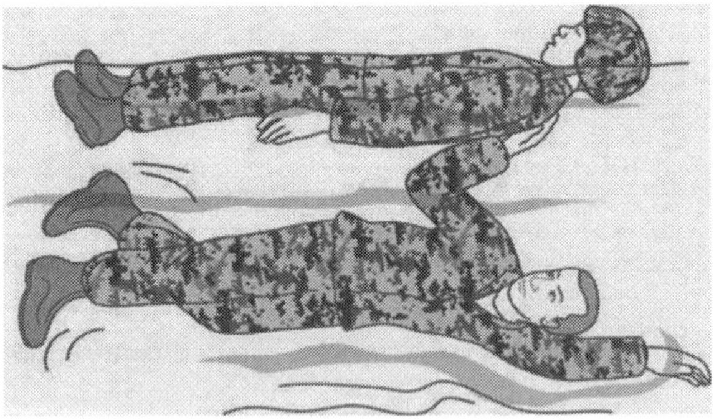

Once the victim is horizontal and on the water's surface, begin forward momentum by extending your towing arm to a fully locked-out position and executing a single armpit tow.

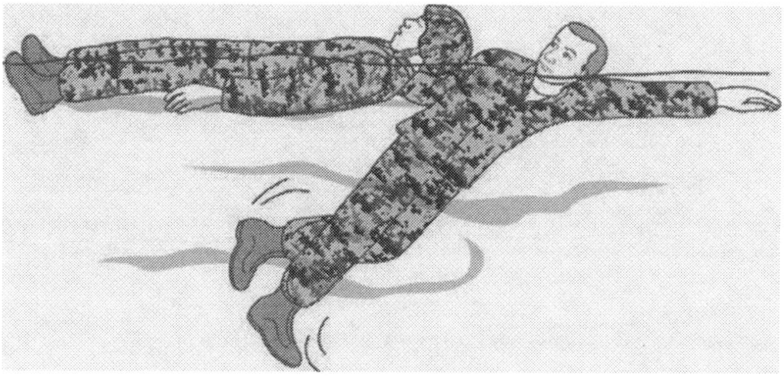

Double Armpit Level Off

Approach the victim.

Stop 2 to 6 yards from the victim to reassess the situation, reassure the victim, and maintain your margin of safety.

Approach the victim slowly and grab his armpits with both your hands. Place your elbows on his back.

Pull with your hands that are in the victim's armpits while pushing with your elbows that are in the victim's back and use an inverted breast stroke kick to place the victim horizontal on the water's surface (face up).

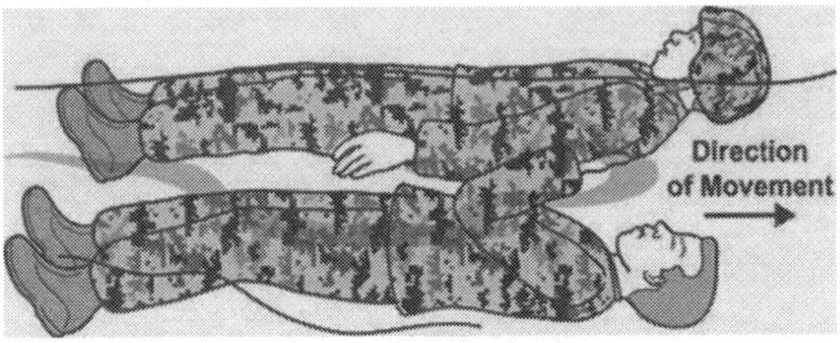

Once the victim is horizontal and on the water's surface, begin rearward momentum by slowly extending your arms into a fully locked-out position and executing a double armpit tow and inverted breast stroke kick.

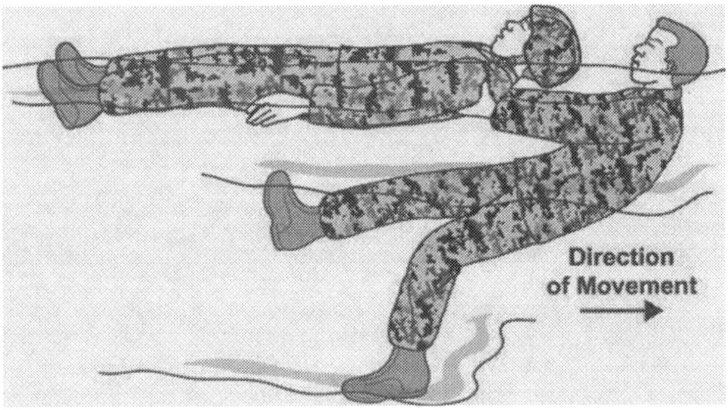

Rescue Techniques

Wrist Tow

Use the wrist tow method to rescue a victim who is floating face down. DO NOT use the wrist tow on a struggling victim. If time allows, remove your helmet and gear before attempting the rescue. Swim toward the victim using a modified breast stroke. Swim within 2 to 6 yards of the victim to maintain a margin of safety, this allows you to reassess the situation and reassure the victim. The following steps show proper front surface approach, wrist tow procedures:

Approach the victim from the front and grasp the underside of the victim's left wrist with your left hand or the right wrist with your right hand. Ensure that your thumb is on the underside of victim's wrist.

Lean back, pulling and kicking strongly to move the victim into a horizontal position.

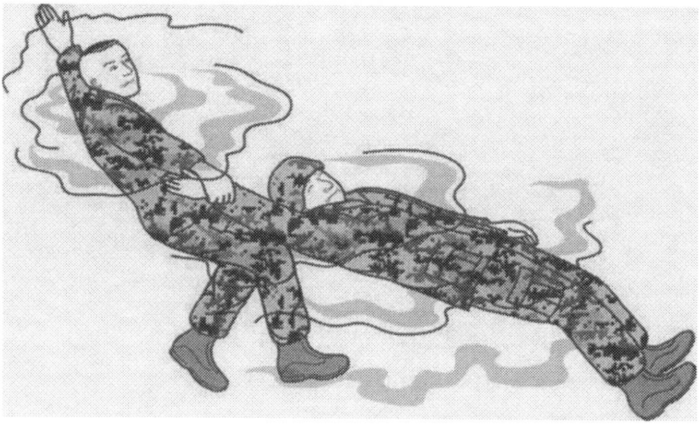

Twist the victim's wrist to rotate the victim into a face-up position.

Swim toward safety using the lifesaving stroke.

> NOTE: The lifesaving stroke is a modified side stroke wherein the top arm is used to tow or carry a victim to safety and use a scissor kick or an inverted scissor kick is used for propulsion.

Keep a firm grip on the victim's wrist.

Keep your towing arm fully extended and along your side. This keeps the victim in the water column and prevents drag.

Ensure the victim's head does not go under water during the recovery.

Single Armpit Tow

You perform the single armpit tow after a victim has been properly leveled off. The single armpit tow uses the lifesaving stroke. To execute the single armpit tow—

Place one hand (top arm) in the victim's armpit. Your towing arm is straight and along your side.

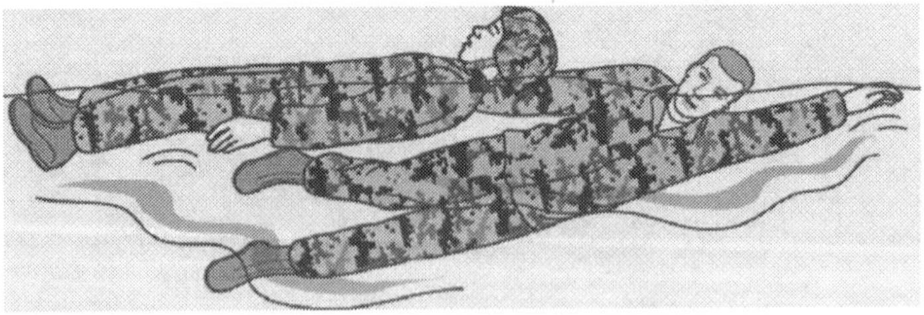

Use your lead arm (bottom arm) to execute short, vigorous pulls.

Execute a scissor or inverted scissor kick in a continuous and vigorous manner.

Use either free breathing or explosive breathing during the tow. During free breathing, keep your head above the water's surface and continuously reassure the victim. During

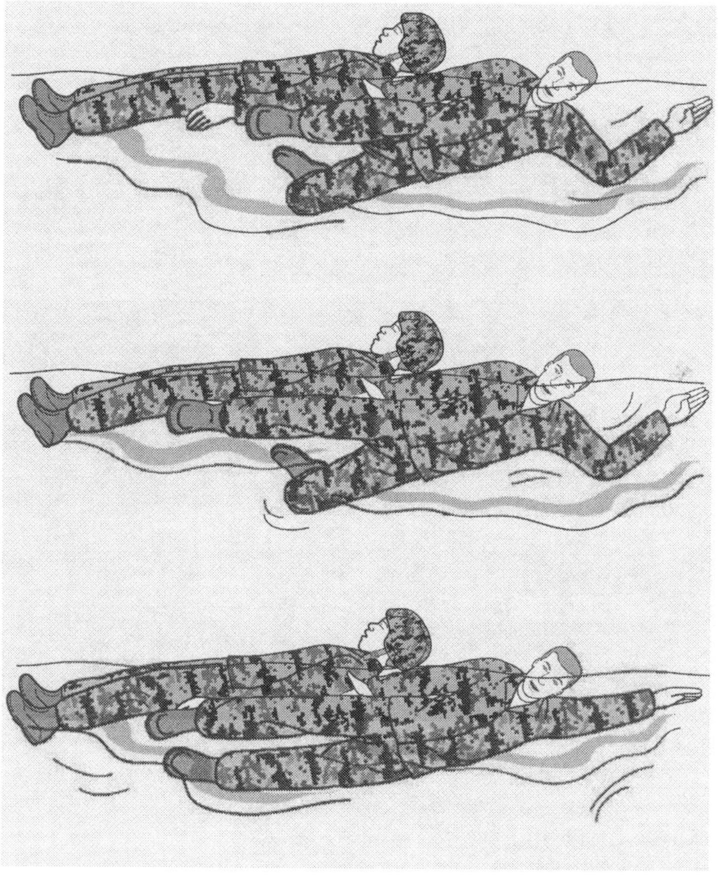

explosive breathing, put your head under the water to plane off your body's angle and reduce drag. Execute approximately two strokes, then lift your head up to breath and reassure the victim. Return your head back underwater for the next few strokes.

Double Armpit Tow

You perform the double armpit tow after a victim has been properly leveled off. To execute the double armpit tow—

Place both hands in the victim's armpits.

Extend both arms fully along your body. You are on your back, and your face is clear of the water.

Use free breathing.

Reassure the victim at all times.

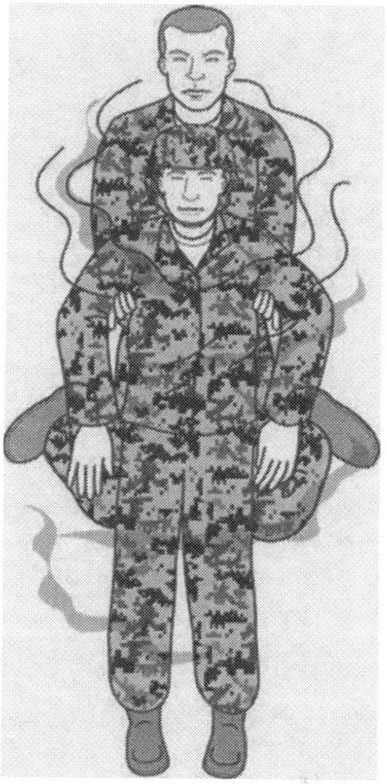

Use the inverted breast stroke kick (it is the only kick that can be used due to your body position) to tow the victim to safety. (The inverted breast stroke kick is the same kick used in the elementary backstroke.) Your kick must be continuous and vigorous in order to keep the victim's face above the water.

Collar Tow

You perform the collar tow after a victim has been properly leveled off. To perform a collar tow using the victim's blouse—

Maintain control of the victim by grasping his armpit with one hand and then with your free hand grasp either his combat gear or his blouse between his shoulder blades. If grasping the blouse, grasp the material with your palm up, then turn your hand over to tighten the material.

Release the victim's armpit once control is established and execute the lifesaving stroke to tow the him to safety.

Cross-Chest Carry

Use the cross-chest carry to carry a victim to safety if the victim is struggling or when moving through heavy surf. Remove your helmet and gear before attempting a rescue. Talk constantly to calm the victim. Swim toward the victim using a breast stroke approach stroke. Swim within 2 to 6 yards of the victim to maintain a margin of safety, this allows you to reassess the situation and reassure the victim. The following steps illustrate proper cross-chest carry procedures:

> **CAUTION:** The cross-chest carry causes fatigue even if you are in excellent physical condition.

Use a level-off technique to place the victim in a horizontal, face-up position.

Retain a grip on the victim with one hand. Reach over the victim with your free hand to encircle the victim's chest. Place your free hand on the victim's opposite rib cage, just below his armpit.

Release your grip once you have a secure hold on the victim's chest.

Swim toward safety using the lifesaving stroke while keeping a firm grip on the victim's chest and your hip on his back.

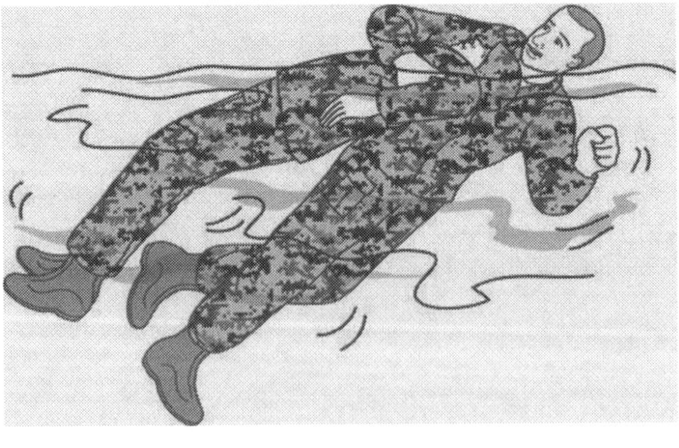

This procedure brings the victim's face and shoulders clear of the water, and typically the victim stops struggling. Sometimes, however, the victim will struggle during the swim to safety. If this happens, either tighten your grip on the victim or defend yourself with one of the techniques discussed on pages 2-25 through 2-30.

Tired Swimmer's Assist

You must maintain a 2 to 6 yards margin of safety from the victim at all times while you are getting into position to perform the tired swimmer's assist. Once behind the victim, you extend one arm, hand straight. Place your straight hand underneath the victim's opposite armpit. Maintain a 45 degree angle away from the victim, your arm is locked out. Assist the victim in propulsion until both of you reach safety.

Defense Against a Drowning Victim

DO NOT sacrifice your life in an attempt to save the victim. A struggling drowning victim poses great danger to anyone nearby. Driven by panic, the victim can grab you with great strength in an effort to climb out of the water. This can result in death for both you and the victim. The following techniques can be used to defend against a drowning victim, but the best defense against attack by the victim is to stay out of reach.

Block

The block prevents the victim from grabbing you if you have approached from the front. If the victim lunges toward you, react as follows:

Place one or both open hands against the victim's upper chest; being careful to avoid the victim's face, neck, and abdomen.

Lean backwards and submerge rapidly. Keep your blocking arm(s) extended.

Swim underwater and away from the victim, and return quickly to the surface.

Stop 2 to 6 yards from the victim to reassess the situation.

Determine an appropriate course of action.

Wrist-Grip Escape/Wrist-Grip Escape Alternative

The wrist-grip escape is used when a victim grabs your arm or wrist. Submerge the victim quickly by reaching across with your free hand, pushing down on the victim's shoulder to submerge him, and kicking to propel yourself upward. While keeping the victim submerged with your hand on his shoulder, give three hard jerking pulls with your trapped hand in an attempt to break free from his grasp. Once free, swim clear of the victim and reassess the victim's condition.

If the wrist-grip escape is unsuccessful in freeing your hand, use the wrist-grip escape alternative: use your free hand to grab your trapped fist, rotate thumbs up, apply bone-on-bone contact with the victim's arm, pry your hand out of his grasp, and quickly swim away from the victim.

Front Head-Hold Escape

The front head-hold escape allows you to escape when you are facing a victim who is gripping you around your head and neck. To execute the front-head hold escape—

Take a quick breath and tuck your chin into your shoulder to protect your throat.

Clap your hands above your head (three times) to submerge instantly. This drags the victim below the water, lifts his arms from around your neck, and, typically, he releases his grasp in order to get back to the surface. If he doesn't release his grasp, apply pressure to the victim's brachial pressure points (which are located inside of the upper arm, above the elbow).

Thrust the victim's arms up and away.

Keep your chin tucked to protect your throat, and swim underwater away from the victim and return quickly to the surface at the ready position.

Stop 2 to 6 yards from the victim to reassess the situation.

Determine an appropriate course of action.

Rear Head-Hold Escape

The rear head-hold escape allows you to escape when a victim is gripping your head and neck from the rear. To execute the rear head-hold escape—

Take a quick breath. and tuck your chin down, turn your head to either side, and raise your shoulders to protect your throat.

Take a strong stroke, clap your hands above your head (three times), and submerge instantly. This drags the victim below the water and, typically, he releases his grasp in order to get back to the surface. If he doesn't release his grasp, apply pressure to the victim's brachial pressure points (which are located inside of the upper arm, above the elbow).

Thrust the victim's arms up and away.

Twist your head and shoulders until free.

Swim underwater away from the victim and return quickly to the surface.

Stop 2 to 6 yards from the victim to assess the situation.

Determine an appropriate course of action.

Administering First Aid/Rescue Breathing

To administer first aid/rescue breathing, remove the unconscious victim from the water when possible and stop any bleeding.

First, open the airway and check for breathing. If the victim is not breathing, give two breaths of air. Check for a pulse. If there is a pulse but the victim is not breathing, continue rescue breathing. If there is no pulse, start cardiopulmonary resuscitation (CPR). For detailed information on administering CPR and rescue breathing, refer to American Red Cross CPR skill cards.

WARNING

If the victim has no pulse and is not breathing, administer CPR immediately. If the victim does have a pulse but is not breathing, give rescue breathing only. If the victim has a pulse and is breathing, DO NOT give CPR—CPR could prove fatal.

Next, protect any wounds from exposure, and treat for shock.

Chapter 3
Treatment of Casualties and Avoidance of Dangerous Marine Life

To survive in the water, you will face many challenges other than just being able to swim. For example, injuries and/or fatigue can lead to drowning and exposure to the elements can lead to hypothermia or heat injuries and may require medical attention. Finally, your presence in the water may be seen as a threat to or food source for a variety of marine life and you may be attacked. This chapter advises you on how to minimize these risks in order to stay alive.

Drowning

Drowning is suffocation by a liquid. A drowning victim inhales water into the lungs or the throat closes by reflex so that little or no water can enter the windpipe. In either case, a victim can no longer breathe.

Symptoms

One symptom of drowning is that the victim may call for help and has an expression of dread or panic. But typically a victim that is active and drowning may not call for help because he is trying to conserve his air and will not speak.

Another symptom of drowning is when the victim thrashes at the water's surface. If the victim stops or grows calmer, he has likely been overcome by fatigue, hypothermia, or a lack of air. At this stage, the victim usually has 20 to 60 seconds before going under the water's surface.

Treatment

If the victim is not breathing, begin rescue breathing. Place the victim on his back, tilt head back to open airway, pinch the nose, and give two full breaths. If the victim does not inhale during the first two breaths, reposition his head and attempt two more breaths. Check for a pulse. If a pulse is present, but the victim is still not breathing, continue rescue breathing. If a pulse is not present, begin CPR. See MCRP 3-02G, *First Aid*, for rescue breathing and CPR details.

WARNING

If the victim has no pulse and is not breathing, administer CPR immediately. If the victim does have a pulse but is not breathing, give rescue breathing only. If the victim has a pulse and is breathing, DO NOT give CPR—CPR could prove fatal. Continue first aid until medical help arrives.

A victim who is not breathing and has no pulse may appear dead. DO NOT decide that death has occurred. Continue with the prescribed treatment. A corpsman or medical officer should decide whether the victim can or cannot be revived.

Hypothermia

Hypothermia is the abnormal lowering of the body's internal (or core) temperature to 95 degrees or below. It occurs when the body loses heat faster than it produces heat. A hypothermia victim loses the ability to move quickly, becomes mentally sluggish, slips into semiconsciousness, lapses into a coma, and dies when internal body temperatures drop too low.

The chilling effects of cold air, wind, or water can produce hypothermia. Water poses the greatest threat because it transfers heat 25 times faster than air. Depending on the water's temperature, a victim can succumb to hypothermia within a few minutes. The body's sudden contact with cold water can also set off a body reaction known as the mammalian diving reflex. This reflex can greatly increase survival time (especially for women and children) in or under cold water. The mammalian diving reflex shuts off blood circulation, except for the flow between the heart, lungs, and brain. The small amount of oxygen left in the blood and lungs is saved for the body's vital organs. This reflex has allowed people to survive being under cold water for an extended period of time. Therefore, a cold water drowning victim should be treated as if still alive even though the victim is not breathing, has no pulse, and may appear dead. DO NOT decide that death has occurred. Continue with the prescribed treatment. Victims of hypothermia can appear to be dead when they are not. A corpsman or medical officer should decide whether the victim can or cannot be revived.

WARNING

If the victim has no pulse and is not breathing, administer CPR immediately. If the victim does have a pulse but is not breathing, give rescue breathing only. If the victim has a pulse and is breathing, DO NOT give CPR—CPR could prove fatal. Continue first aid until medical help arrives. Check for a pulse for at least 45 seconds.

Symptoms

Once the body's core temperature drops, the victim will show one or more of the following symptoms:

- Violent and uncontrollable shivering as the body tries to warm itself.
- Slow or slurred speech.
- Disorientation or poor coordination.
- Loss of skin color.
- Blue and pinched lips.
- A slowing or stopping of shivering that progresses into a rigid torso and limbs.

Survival Time

A hypothermia victim's survival depends on the water's temperature and the time spent in the water. A small body build cools faster than a large build. Children cool faster than adults.

To increase your chance of survival in the water, utilize the HELP position described on page 1-37. Extra clothing and inactivity (remaining motionless in the water) can also increase your survival time.

Treatment

A hypothermia victim must be warmed to prevent further heat loss; therefore, treatment should begin as soon as possible. Consciousness of the victim determines the treatment that should be pursued. See MCRP 3-02G for specific treatment information. The following treatment procedures are recommended:

- If the victim is conscious, give the victim warm fluids. Give candy or sweetened foods to a victim who is able to eat.
- If the victim is unconscious, place him on his back with his head tilted back to ensure an open airway.
- DO NOT massage the victim. Massage can break blood vessels and create swelling, internal pressure, and blocked blood circulation.
- DO NOT give alcohol to the victim. Alcohol lowers the victim's body temperature.
- Shock is a possibility, treat accordingly.
- Seek medical help immediately.

If you are able to remove the victim from the water, apply the following steps when possible:

- Get the victim into shelter.
- Remove the victim's wet clothing.
- Put the victim in dry clothing.
- Place the victim in a sleeping bag if one is available. It may be necessary to place another Marine in the sleeping bag with the victim.
- Place as much insulation as possible between the victim and the ground.
- Use hot water bottles, electric blankets, or blankets heated in an oven or by a campfire to warm the victim's neck, groin, and the sides of the chest.

CAUTION: DO NOT apply heat to extremities.

Heat-Related Injuries

Heat-related injuries include heat cramps, heat exhaustion, and/or heat stroke. They can occur when a Marine—

- Is exposed to extreme heat, such as from the sun or a combination of high air temperatures and water temperatures.
- Does not wear proper clothing or gets overexposed to the sun.
- Becomes dehydrated.

With heat cramps, muscles may cramp in the arms, legs, and/or stomach. The victim may sweat excessively. To treat, create improvised shade for the victim and have him drink water.

With heat exhaustion, the victim sweats heavily; presents pale, moist, cool skin; and complains of a headache, weakness, dizziness, and/or loss of appetite. A victim may also experience heat cramps, nausea, vomiting, an urge to defecate, chills, rapid breathing, confusion, and a tingling sensation in the hands or feet. To treat, pour water over the victim and fan him to speed up the coolant effect of evaporation, have the victim drink freshwater, and attempt to provide shade.

With heat stroke, the victim stops sweating, which results in red, flushed, hot, dry skin. The victim may first experience headaches, dizziness, nausea, fast pulse and respiration, and/or seizures and mental confusion. The victim can become unconscious and die if not treated quickly. To treat, create improvised shade for the victim. Have him drink water and elevate both legs if possible.

Burns

Apply a field dressing (or the cleanest material available) to the burn. Give sips of water to a casualty who is conscious and not nauseated. When treating burn casualties—

- DO NOT remove clothing stuck to the burns.
- DO NOT break any blisters.
- DO NOT apply grease or ointments to the burns.

For electrical burns, check for both an entry and exit burn from the passage of electricity through the body. An exit burn may appear on any area of the body, not necessarily opposite the entry burn.

For burns caused by wet or dry chemicals, flush the burns with large amounts of water and cover with a dry dressing.

For burns caused by white phosphorus, flush the area with water, then cover with a wet material, dressing, or mud to exclude air and keep the white phosphorus particles from burning.

For laser burns, apply a field dressing.

Common Medical Problems Associated with Sea Survival

Seasickness

Seasickness is the nausea and vomiting caused by the bobbing motion created by the wave action of a flotation device. Seasickness can result in—

- Dehydration and exhaustion.
- A loss of the will to survive.
- Others becoming seasick.
- Unclean conditions.

To treat seasickness—

- Wash both the Marine and the flotation device to remove the sight and odor of vomit.
- Keep the Marine from eating food until the nausea is gone.
- Have the Marine lie down and rest.
- Give the Marine seasickness pills if available. If the Marine is unable to take them orally, the pills should be inserted rectally for absorption by the body.

Saltwater Sores

Saltwater sores occur when skin that has abrasions or is cut is exposed to saltwater. The sores may form scabs and pus. Do not open or drain. Flush the sores with freshwater, if available, and allow to dry. Apply antiseptic, if available.

Blindness/Headache

Irritants or the effects of the sun's rays reflecting off the water can cause temporary blindness or headaches. If flames, smoke, or other irritants get in your eyes, flush the eyes immediately with saltwater, then with freshwater, if available. Apply an ointment, if available. Bandage both eyes for 18 to 24 hours or longer if the damage is severe. If glare from the sky and water causes your eyes to become bloodshot and inflamed, bandage the eyes lightly. Try to prevent this problem by wearing sunglasses or goggles with a sunglass insert.

Constipation

This condition is a common problem associated with dehydration. For constipation, do not take a laxative if it is available; this causes further dehydration. Drink freshwater, if available.

Sunburn and Dehydration

The sun's rays reflect at all angles off the waves of the water; therefore, sunburn and dehydration are serious problems in sea survival. Try to prevent sunburn by—

- Erecting an improvised canopy, with available floating materials, to provide shade.
- Wearing your soft cover or using a cloth, such as a handkerchief, to cover your head.
- Covering your skin with sunscreen or lip balm from your first aid kit. Your lips, nostrils, eyelids, the backs of your ears, and the skin under your chin sunburns easily. If enough sunscreen cream is available, all exposed skin should be covered.

Dehydration is caused by the loss of the body's vital fluids. Dehydration in saltwater may result from a combination of factors such as a lack of water, the effects of saltwater on skin tissue, sunburn or vomiting from seasickness, and other causes. Sleep and rest and reduced water and food intake are the best ways of enduring periods of exposure. The following measures will delay the effects of dehydration:

- DO NOT drink saltwater.
- DO NOT drink urine.
- DO NOT drink alcohol.
- DO NOT smoke.
- DO NOT EAT unless water is available.

Dangerous Marine Life

You may see many types of marine life around you, and some are more dangerous than others. Generally, sharks are your greatest danger, followed by barracudas. However, most marine life will not deliberately attack a human. The most common injuries from marine life are wounds from bites, stings, or punctures. With the exception of sharks and barracudas, most injuries are a result of either trying to catch game or from contact abrasion with marine life. To treat an injury that results from a dangerous marine animal bite—

- Pack the wound with gauze and then apply a pressure bandage, if available.
- Treat for shock.
- Obtain medical attention as soon as possible.

For a more detailed information on the treatment of casualties and dangerous marine life, see MCRP 3-02F and MCRP 3-02G.

Sharks and Barracudas

Only about 20 percent of all shark species are known to attack people. Sharks have an acute sense of smell, and the smell of blood in the water will draw them to their prey. They are also very sensitive to any abnormal vibrations in the water; therefore, the sound caused by a struggling swimmer or underwater explosions will attract them.

Barracudas are bold and inquisitive fish. They have been known to attack men who are wearing shiny objects. Barracudas may charge at lights or shiny objects at night.

A group of swimmers can maintain a 360 degree watch while in the water. Therefore, to protect yourself from sharks and barracudas, stay with other swimmers. The group members can either frighten away or fight off sharks or barracudas better than one person can. Keep all your clothing on, including footwear. Historically, sharks attack unclothed individuals in groups first, mainly in the feet. Clothing also protects you against abrasions from a shark's tough skin should the shark brush up against you. Avoid urinating heavily, let urine dissipate between discharges. If you must defecate, do so in small amounts and toss it as far away from the group as possible. Do the same if you must vomit.

If attacked, the use of firearms by swim sentries should be used with extreme caution because of the risk of injury to other swimmers. If unarmed or unable to make an improvised weapon, kick and strike the shark. Avoid using your bare hands to strike the shark, injury can result to your hands due to a shark's tough skin. Target areas on a shark are the gills, eyes, and underbelly. Blows to the snout are also not recommended because a shark will tilt its head up and thrust its jaws forward when biting.

Sea Snakes

Sea snakes are venomous and sometimes found in mid-ocean. They are unlikely to bite unless provoked. Your best protective measure is to avoid them.

Poisonous Fish

Many reef fish are toxic, and can kill you if eaten. Generally, poisons are present in all parts of the fish, but especially concentrated in the liver, intestines, and eggs.

Turtles and Moray Eels

Turtles and moray eels normally inflict minor bite wounds. Treat this type of injury by cleaning the wound.

Corals

Coral, dead or alive, can inflict painful cuts. Clean all coral cuts thoroughly. DO NOT use iodine to disinfect any coral cuts. Some coral polyps feed on iodine and may grow inside the flesh.

Jellyfish, Portuguese Man-of-War, Anemones, and Others

This group of marine animals inflicts injury by stinging their victims with their tentacles. Contact with their tentacles produces burning pain, a rash, and small hemorrhages on the skin. Shock, muscular cramping, nausea, vomiting, and respiratory distress may also occur. Gently remove the clinging tentacles with a towel and wash or treat the area. Use diluted ammonia or alcohol and talcum powder to treat the injury if available.

Spiny Fish, Urchins, Stingrays, and Cone Shells

These animals inject their venom by puncturing the skin with their spines. General signs and symptoms include swelling, nausea, vomiting, generalized cramps, diarrhea, muscular paralysis, and shock. Deaths are rare. Treatment consists of soaking wounds in hot water, if available, to deactivate heat-sensitive toxins.

Chapter 4
Negotiating Water Obstacles

Marines face water obstacles in saltwater, freshwater, and brackish water (where freshwater and saltwater meet). These water environments differ considerably and pose distinct problems for Marine tactical units and swimmers. Saltwater obstacles include tides, surfs, and currents. Brackish water obstacles include back bays. Freshwater obstacles include rivers and canals.

Tides

Tides are periodic changes in the surface levels of oceans, bays, gulfs, inlets, and rivers. The Moon's and Sun's gravitational pulls cause tides. Depending on the situation, tides can either help or hinder Marines in their endeavors to conduct amphibious or riverine operations. Tides can provide sufficient depths to allow for the passage of landing craft or boats over reefs, trees, and other underwater obstructions. On the other hand, tides can render a river fording site unusable to tactical vehicles. Direction, level of change, and amount of change determine tidal nomenclature.

Tides that show change in direction are flood tides and ebb tides. Rising tides are known as flood tides. Falling tides are known as ebb tides.

Tides that show extreme levels of change are high tides and low tides. High tide is the period when water is at its greatest depth. Low tide is the period when water is at its most shallow depth.

Tides that show the least and most amount of change are neap tides and spring tides, respectively. Neap tides have the least amount of change in water levels between high and low tide. Neap tides occur at the half moon, when the Sun and the Moon are aligned at a 90-degree angle with the Earth. In this position, the Sun's and Moon's gravitational pulls offset each other. Spring tides have the highest floods and lowest ebbs. Spring tides occur at or shortly after the new moon or full moon when the Sun, Moon, and Earth are approximately in line. In this position, the Sun's and Moon's gravitational pulls are combined.

Surf

Waves break upon entering shallow water and create surf. The offshore area where waves break is the surf zone, which can present many hazards. Breaking waves often trap air bubbles and create a foamy appearance. Bubbles lower the water's density and decrease buoyancy. Move through foamy surf as quickly as possible. The type of wave determines your survival technique.

Wave action moves you toward shore. Lie on your back or side with your head pointing in the direction of the beach and your feet pointing into the waves.

As one wave approaches the beach, another drains away; relax and do not swim against the draining water. When a new wave is within about 3 yards, start swimming toward shore. Continue to swim until the wave lifts you and moves you toward the beach. Once the wave loses forward momentum, relax and repeat the cycle. If nearing rocks, turn your body and approach feet first to reduce the chance of striking your head and arms.

Plunging Waves

A plunging wave is a breaker whose crest curls forward and falls ahead of its base. Because of its power and underwater turbulence, a plunging wave poses the greatest surf threat. If caught in a plunging wave, you can be pulled underwater and pitched about violently. This can cause you to panic, which can increase your chance of drowning. Perform the following steps to escape a plunging wave:

Tuck into a ball with your head against your knees and your forearms locked around your legs, just below the knees.

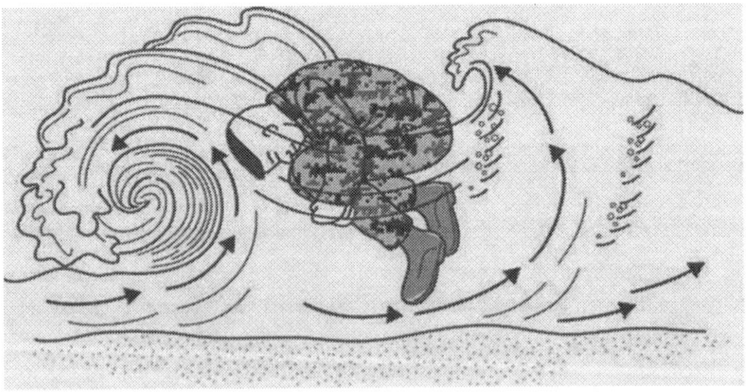

Relax in this position until the turbulence subsides and you float to the surface. This can take 30 seconds or more.

Swim toward shore.

NOTE: If threatened by another plunging wave, dive underwater into the wave.

Spilling Waves

A spilling wave does not break. Instead, its crest slides forward without curling. A spilling wave creates less turbulence and poses less of a threat than a plunging wave. If caught in a spilling wave, relax and let the wave carry you to shore.

Surging Waves

A surging wave occurs on a beach with a steep underwater gradient. It never really breaks, but the crest rises while the base slides up the beach with great force and speed. Once the wave reaches its highest point on the beach, it rushes back as quickly as it surged forward. If you are standing on the bottom when a surging wave advances or retreats, the wave can knock you off your feet and pull you into the surf zone. Do not try to stand or walk on the bottom. Swim toward the beach as soon as possible.

Currents

Offshore Currents

An offshore current occurs outside the surf zone. Typically, it occurs at bay entrances, in island channels, and between islands and the mainland. An offshore current flows parallel to or away from shore. If the offshore current is created by tides, its current strength and direction vary at different times of the day.

If caught in an offshore current, you may be carried in a direction you do not want to go. DO NOT try to swim directly to safety. If the current is moving directly away from the shore, relax and wait until the current dies out or turns toward land. Once the current subsides, use any survival stroke to swim toward shore. If the current is moving parallel to shore, use any stroke to move at an angle across the current and toward shore.

Rip Currents

A rip current occurs when waves pile water against the shore faster than the water can drain. The water flows rapidly along the beach until it is deflected seaward by a bottom obstruction. Then the rip current flows through the surf zone and into open water at a speed of up to 2 knots. This action can cut deep trenches in the sand. A rip current dies out once it reaches open water (usually within a few hundred yards of the shore).

A rip current can pull you out to the open sea. If caught in a rip current, DO NOT try to swim against the current. A rip current moves faster than most people swim, and it is impossible to swim to shore once caught in it. Relax and stay afloat until the current runs out. Once the current subsides, use a survival stroke to move parallel to the shore until you are out of the current, then begin swimming toward shore.

Littoral Currents

A littoral current occurs when a wave breaks against a beach at an angle. This current flows parallel to the shoreline and does not pose a great threat. If caught in a littoral current, use the combat travel stroke to swim across it at an oblique angle.

Back Bays

Once on the beach, you often face one or more rows of low hills called dunes. Behind the dunes, you may encounter a low-lying stretch of ground thickly covered with scrub trees and bushes. This area gives way to wetlands known as back bays. Back bays consist of muddy islands that are almost submerged during flood tides and separated by channels of brackish water of varying depths. Channel bottoms usually contain soft mud. Back bays pose major obstacles to vehicular traffic.

Infantry can cross back bays, but only with great effort. If crossing back bays by foot, consult detailed navigation charts and use the following guidelines to plan your route:

- Avoid water less than waist deep; walking in shallow water or soft mud is extremely tiring.
- Avoid back bay islands; these low-lying islands are usually too muddy to support foot traffic.
- Seek out deep water; floating with a pack is less tiring than walking through shallow water or soft mud.
- Seek out sand, shell, gravel, or stone bottoms; these firmer bottoms generally ease travel and help conserve energy.

Rivers and Canals

A river is a large, natural stream of water that empties into a larger body of water. The slope of the riverbed and the volume of water in the river determine its current.

Canals resemble small rivers or streams in their width and depth, but usually lack any significant current. Climbing out of these waterways can be difficult if the canal is flanked by steep banks.

Chapter 5
Fording Waterways

WARNING

Fords are dangerous. Cross as quickly as possible.

A ford is any site in a river, stream, or canal where the water is shallow enough for troops or vehicles to cross without using flotation devices. Canal bottoms are usually too soft to support fording vehicles, and wading infantry frequently stumble.

The tactical situation dictates the location of the fording site. Seek fords that are protected from enemy observation and that allow for adequate supporting fires. A night fording takes at least one and a half times as long as a daylight fording.

> *NOTE:* In this chapter, the term "river" refers to rivers, streams, and canals.

Selection of a Ford Site

The following table identifies desirable fording characteristics:

Characteristics	Comments
Concealment	The ford hides personnel and vehicle movement from enemy observation.
Accessibility	The ford should have low banks with gentle gradients. This allows a free flow of traffic at both the entrance and the exit.

Characteristics (Continued)	Comments (Continued)
Slow Current	The ford's current should not exceed 1.5 meters per second if possible.
Firm Footing	The ford's bottom, entry, and exit composition should be firm enough to support traffic. Do not drive a vehicle over any bottom composition that a 2-inch diameter stick can be pressed into more than 1 or 2 inches.
Gently Sloped Channels	The ford's entry and exit points should be gently sloped. If possible, locate a portion of the stream where the channel is not actively shifting.
Depth	The fording depth is less than or equal to the least capable vehicle.

Determine the Slope

Units move into and out of water faster and more quietly if entry and exit points are not steep or muddy. Slope is the amount of change in ground horizontal distance (run) and in vertical elevation (rise) from one point to another. Slope is usually expressed as a percentage. You can use a clinometer, map, or line of sight and pace to measure the percentage of a slope.

Clinometer

A clinometer measures percentage of a slope. It is a component of the M9 armored combat excavator, and organic to many Marine engineer units. However, engineers are also taught (and often use) field-expedient methods to determine slopes and gradients.

Map

A map measures the horizontal distance along a desired path. Determine the difference in elevation between the path's starting and ending points. From a map's scale, you can determine the distance between two points. From a map's elevation lines, you can determine the difference in height between the same two points. Both figures must be in the same unit of measure (e.g., feet, meter). Divide the elevation (rise) by the distance (run) and multiply by 100.

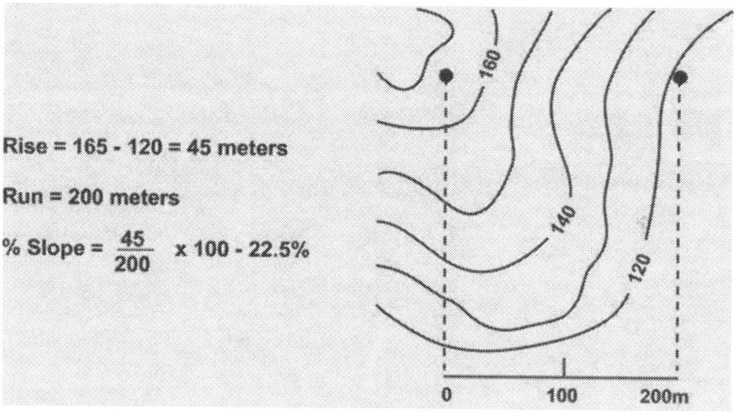

Line of Sight and Pace

To determine line of site and pace, stand at the bottom of the slope, keep your eyes level, pick a spot on the slope, then pace the distance. The number of paces multiplied by a standard measure of 0.75 meter determines the run. The eye-level height (usually 1.5 to 1.7 meters) determines the rise. Repeat this procedure until you have covered the entire distance you want to measure for each spot (vertical and horizontal). Add the vertical distances to provide total rise and the horizontal distances to provide total run.

Determine the Current Speed

Current speed increases as channels narrow. It may be necessary to locate a wider ford location to obtain a slower stream current. The following steps are used to calculate the speed of the current:

Determine points A and B along a channel. Then measure the distance between those two points.

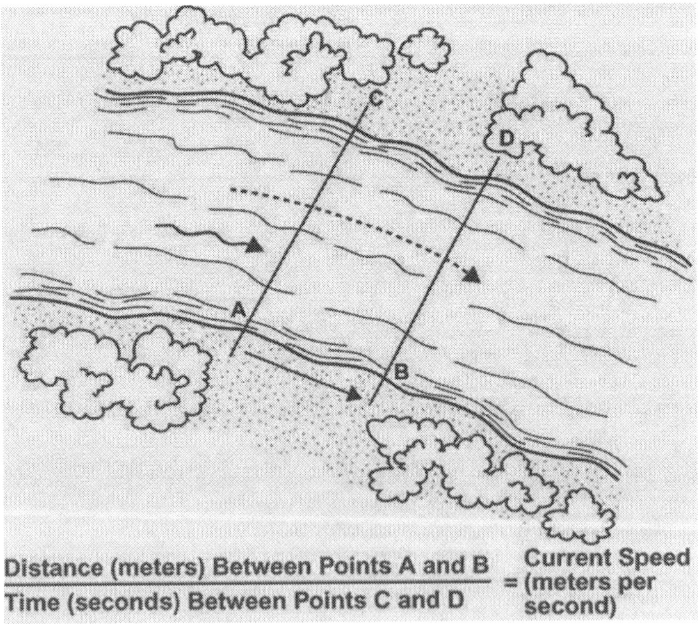

$$\frac{\text{Distance (meters) Between Points A and B}}{\text{Time (seconds) Between Points C and D}} = \text{Current Speed (meters per second)}$$

Sight directly across the water from points A and B to locate points C and D.

Throw a floating object (e.g., a stick) upstream from points A and C. Observe the object as it floats toward points B and D.

Subtract to find the time it takes for the object to float from start to finish.

Do not attempt to swim across currents that are moving faster than 1.5 meters per second. Equivalents of this speed include—

- Quick-time march rate of 120 counts per minute with one, 30-inch step at each count.
- 5 feet per second.
- 3.5 miles per hour.
- 5.5 kilometers per hour.

Measure River Width

A river's width can be estimated from the width of its symbol on a scaled topographic map. If this is not possible, use the following compass techniques:

Stand at the water line (A).

Shoot an azimuth to a point on the opposite bank (B).

Move upstream or downstream until you are at a point (C) where you can shoot an azimuth 45 degrees larger or smaller than the original azimuth.

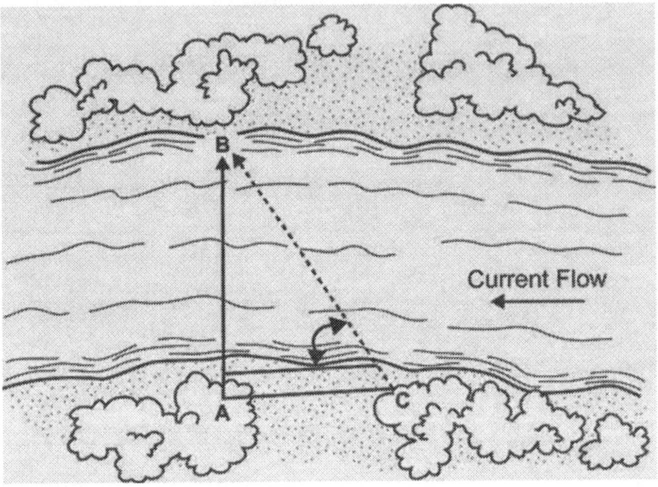

Measure the distance between points A and C. The distance calculated equals the river's width.

Calculate Downstream Drift

A river's current causes personnel and equipment to drift downstream. If personnel and equipment are aimed straight across the river, they will sideslip downstream as they move across the current to the other shore. Therefore, personnel and equipment crossings must compensate for the effects of a river's current and entries are usually made upstream of the desired exit point. Use the formula below to calculate downstream drift.

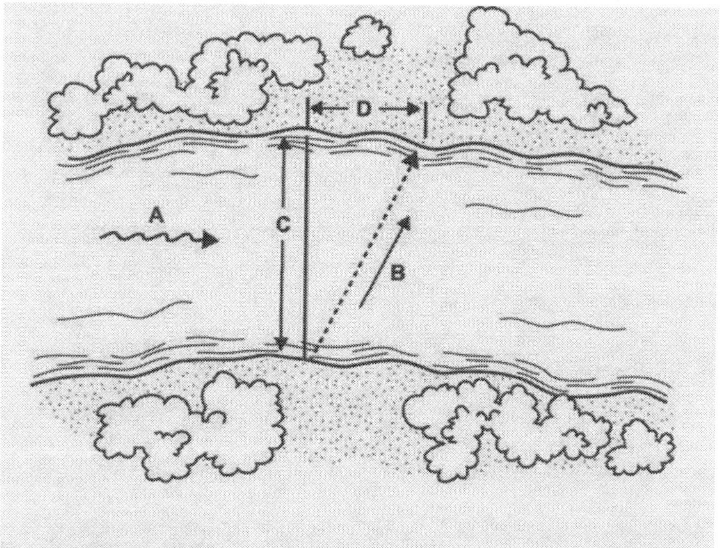

$$\frac{\text{Current Speed (A)}}{\text{Crossing Speed (B)}} \times \text{River Width (C)} = \text{Downstream Drift (D)}$$

> *Note:* The crossing speed for a swimmer across a river may vary but is generally limited to 1 meter per second. All measurements must be in the same unit of measure (e.g., meters, feet).

The Buddy System

Whenever a Marine unit must enter into or operate on the water, a "buddy system" is employed in which every Marine is paired with a swimming partner. The buddy system matches an experienced swimmer with a weak swimmer. The experienced swimmer assists and encourages the weaker swimmer and bolsters confidence during night crossings. If a unit has an odd number of Marines, place the extra person with another pair to form a three-person team.

Water Crossings

Care of Weapons

Marine infantry weapons and munitions are designed to be able to operate after immersion. However, protect your weapons from moisture whenever possible.

A gas-operated weapon can malfunction if water travels down the barrel and enters the gas tube. To protect the gas tube—

Close the weapon's bolt before entering the water.

Seal the muzzle with a condom, balloon, plastic spoon wrapper, or other form of waterproof material.

Tie or melt the protective cover to create a watertight seal.

When the muzzle's protective cover is no longer needed, remove it. Open the bolt and inspect the barrel. If the tactical situation permits, swab excess moisture from the barrel. Test fire automatic weapons, if possible. Field strip and clean weapons as soon as possible. If time does not allow for a complete inspection, rinse inaccessible areas with small amounts of diesel fuel, then dry.

ISOMAT Raft

Construction of an ISOMAT raft is time-consuming. This type of raft should not be employed as part of an attack, but used for logistical purposes (e.g., evacuating stretcher cases, transporting supplies). Use the following steps to build an ISOMAT raft:

Wrap ISOMAT sleeping pads around sturdy sticks.

Use parachute cord and square knots to tie the pads securely in place and to lash stick ends together in a rectangle.

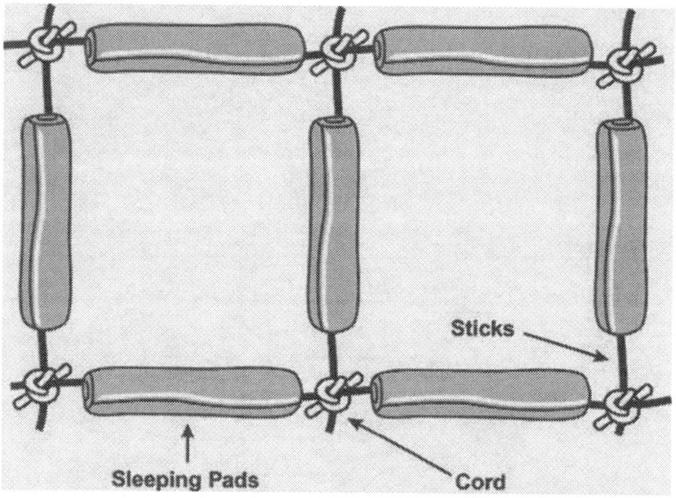

NOTE: The ISOMAT raft pictured can support several hundred pounds. However, the cargo will get wet if not properly waterproofed.

Poncho Raft

A poncho raft can support two Marines and their equipment and is well suited for long crossings. Use the following steps to build a poncho raft:

Inspect two ponchos and ensure they are serviceable.

Lay one poncho flat on the ground, with the hood-side up.

Cinch the hood tightly to form a gooseneck or tie in a knot.

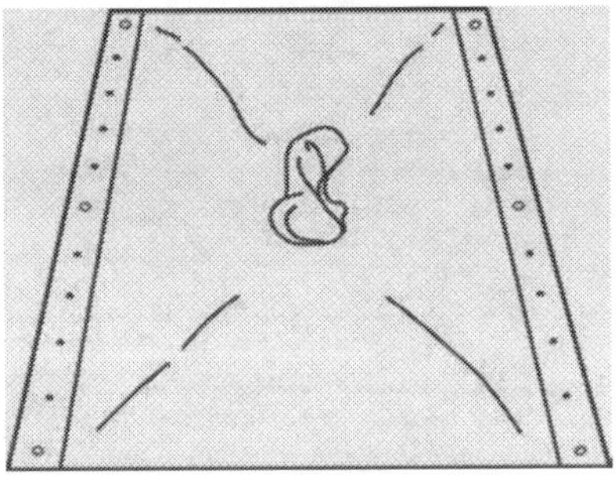

Pad sharp edges of equipment and place the equipment in the center of the poncho.

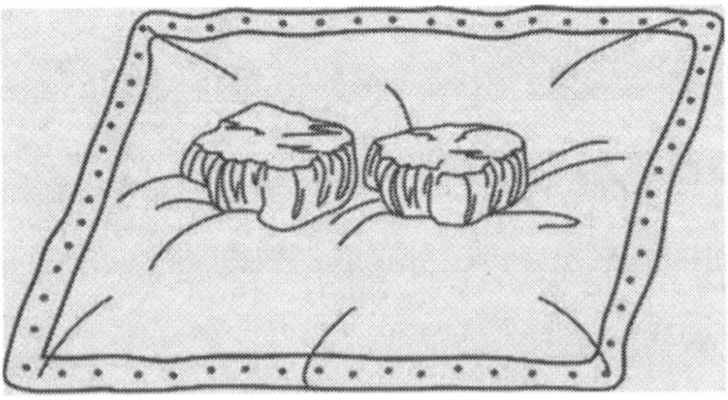

Place the second poncho over the equipment, rubber side up, and hood facing down.

Snap the edges of the two ponchos together.

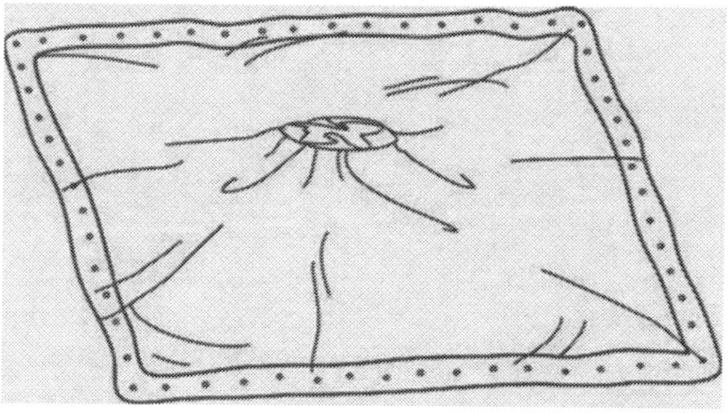

Roll the edges toward the equipment.

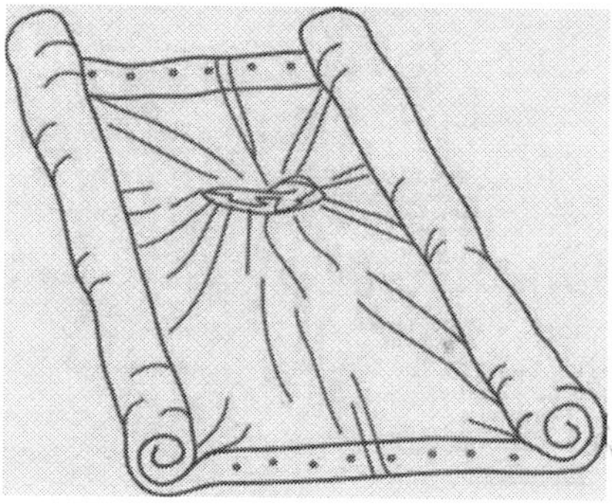

Roll the edges into pigtails and tie them off.

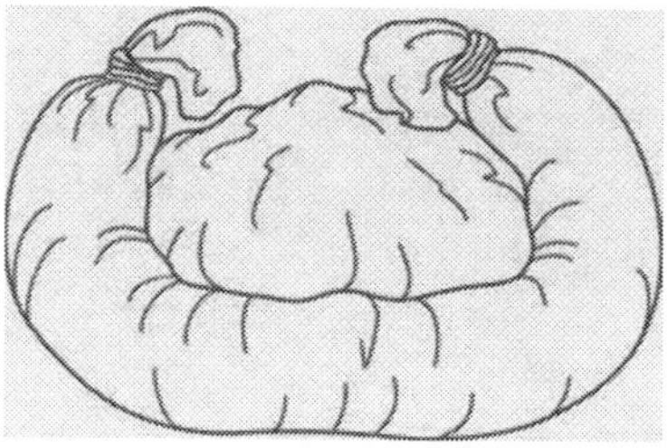

Pull the pigtails together over the top and lash them securely.

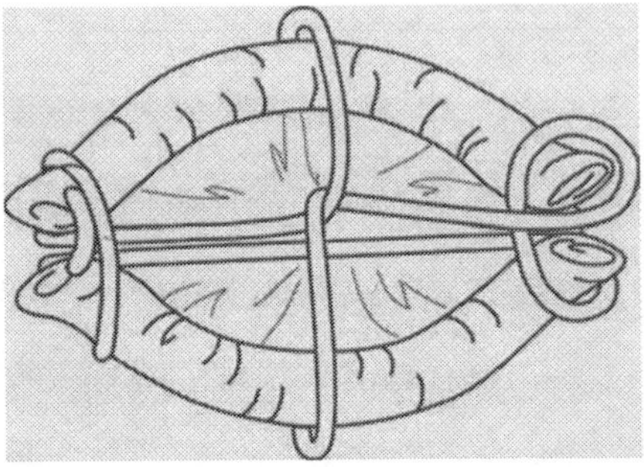

Protect the raft from brush punctures while placing it in the water. Swim across the water obstacle while security elements are covering the far shore.

Construction of a Pack Raft

You will need two waterproofed packs and two M16A2 service rifles. The following steps are required to construct a pack raft:

Place two packs side-by-side with the pack frames on the deck. The tops of the packs are opposite of each other.

Loosen the main compartment straps on both packs.

Insert one rifle on each end between the straps and the packs. The muzzles are opposite of each other. The rifles serve as one means to secure the packs together. Place the front sight post under the top flap.

Tighten the straps so that the rifles and packs are secure.

Take the excess strap on the inner side of each pack and secure it to the opposite pack to better secure the two packs together.

Take the excess straps on the outer sides of the packs and use those straps as safety lashing for the rifles.

Tuck the excess straps and check to make sure the rifles and packs are secure.

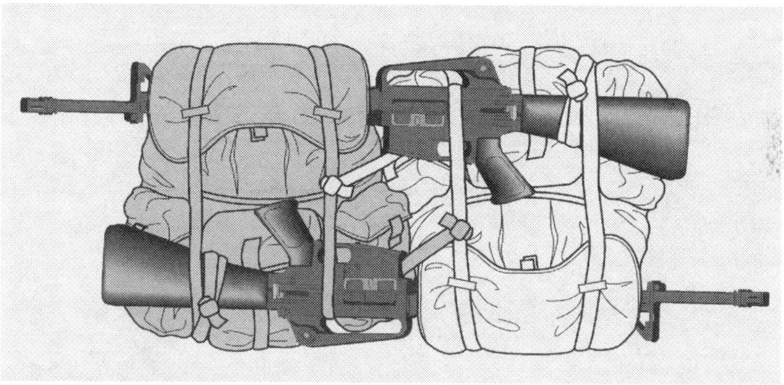

Single-Rope Bridge

A single-rope bridge offers a temporary and quick way to cross small rivers. It also provides extra security while crossing swift waters. At night, it prevents straggling, and guides units precisely from one side of the river to the other side. If crossing a river at night, plan for at least one single-rope bridge.

If your unit is crossing a river with swift currents or water depths above 4 feet, the unit is carrying sufficient rope to span the crossing site, and the tactical situation permits, secure the rope on near and far banks to provide a hand-hold for crossing Marines. This reduces the time required for the entire unit to cross and provides a degree of comfort/confidence for poor swimmers. Use a squad-sized bridge team to construct a single-rope bridge. Station several strong swimmers at the water's edge to help anyone who has trouble crossing.

Nylon rope is normally coiled in 120 foot lengths. It is 0.6 inches in diameter and has a breaking strength of about 3,840 pounds. Over time, a nylon rope can stretch to as much as one-third more than its original length and stretching weakens the rope. If the rope is stretched, discard the rope or use it for light tasks. To prolong the life of a nylon rope, do not step on it or drag it on the ground. Pad the rope in places where it contacts rocks or sharp corners. Do not leave the rope knotted or stretched longer than necessary. Dry rope as soon as possible. Single-rope bridge construction is as follows:

Tie a sling rope around your waist using a square knot and two, separate half hitches. See the appendix for detailed information on knots.

Attach a locking steel carabiner to the sling rope.

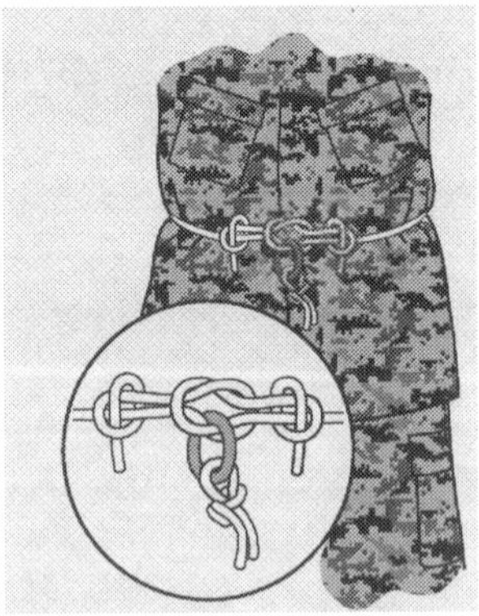

Tie a bowline knot in the running end of the bridge rope and attach it to the carabiner.

Temporarily secure the other end of the rope to a tree on the near shore.

Enter and cross the water.

NOTE: Carry only your weapon and ammunition.

Exit the water on the opposite shore.

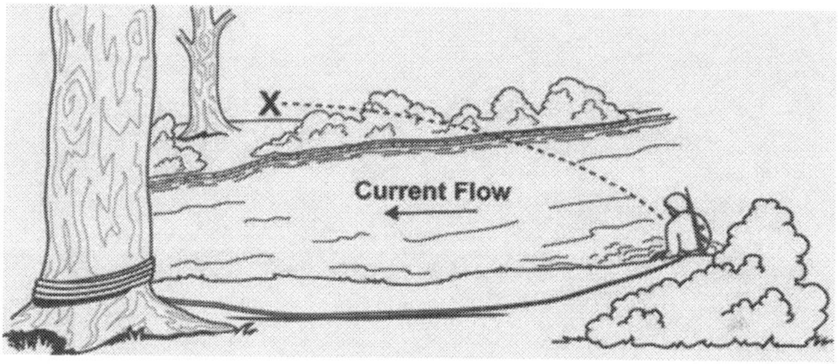

Prepare your weapon for use. Unhook the bridge rope from the carabiner at your waist, and tie the bridge rope to a sturdy tree using a round turn and two half hitches.

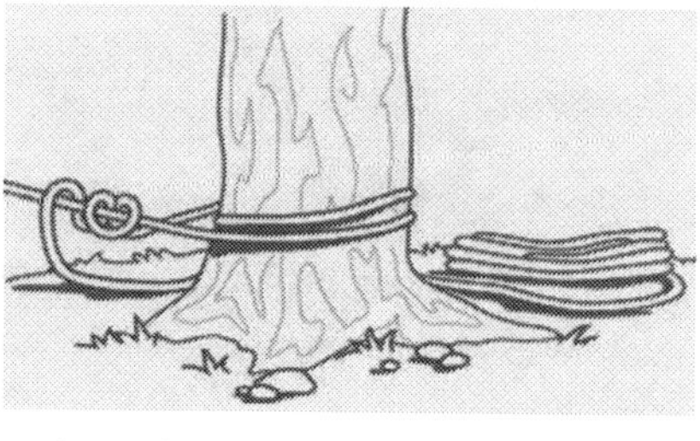

Conduct a box reconnaissance of the opposite shore.

On the near shore, have another Marine prepare to tighten the rope. That Marine should place a transport tightening system in the bridge rope by tying a double butterfly knot and placing two carabiners in the butterfly. See the appendix for detailed information on knots.

The Marine should pass the running end of the bridge rope around the downstream side of the near shore anchor point and through the two carabiners.

Pull the butterfly knot approximately one-third of the distance across the river.

Secure the bridge rope to an anchor point using a round turn and two half-hitches.

On the near shore, the Marine helping you should pull the slack out of the bridge rope until the butterfly knot is back on the near side. The bridge rope is then tied off against itself using two half hitches with a quick release in the last half hitch.

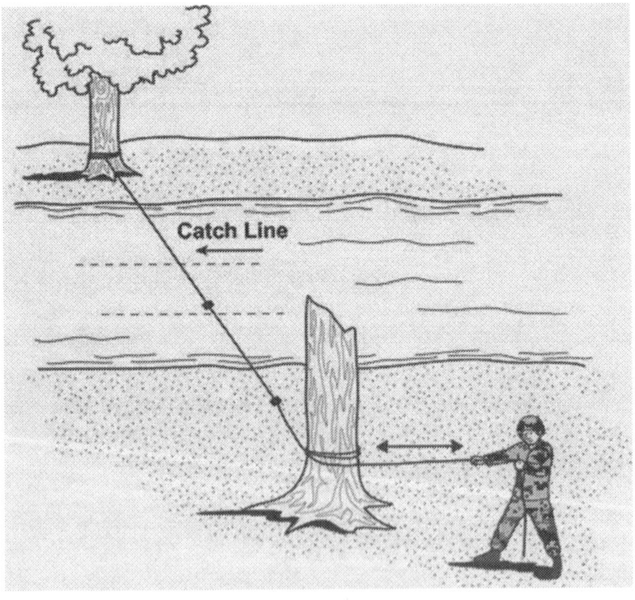

NOTE: The single-rope bridge must be as tight as possible so it will not sag when used.

If you lose your footing and fall into the water, swim with the current to the closest shore. Swimming against the current is dangerous and quickly causes fatigue.

High and Dry Crossings

If the single-rope bridge is high enough, suspend yourself below the single-rope bridge and above the water. Use the following steps to suspend yourself from a single-rope bridge and then pull yourself across the water:

Tie a sling rope around your waist using a bowline. Ensure that the knot is tight.

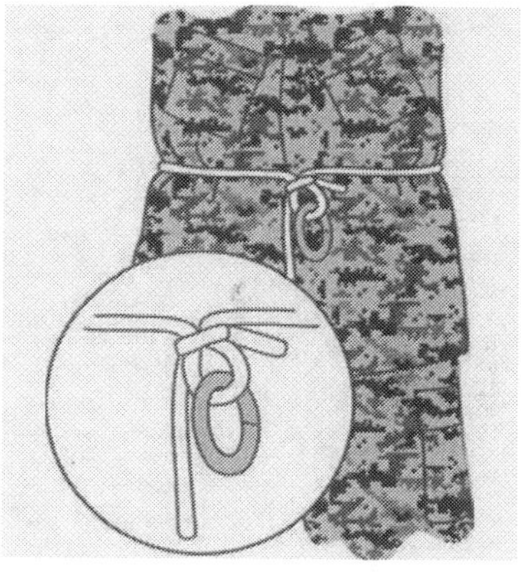

Attach a carabiner through the bowline's loop. The carabiner's gate faces up.

Secure your helmet chin strap.

Face the single-rope bridge with your left shoulder toward the far shore.

Grasp the bridge rope in both hands.

Swing your body beneath the single-rope bridge with your head toward the far shore. Cross your ankles above the bridge rope.

Arch back until the carabiner contacts the bridge rope. Connect the carabiner to the bridge rope. Allow the carabiner to bear your body's weight.

Pull yourself across the single-rope bridge, hand over hand, to the far shore.

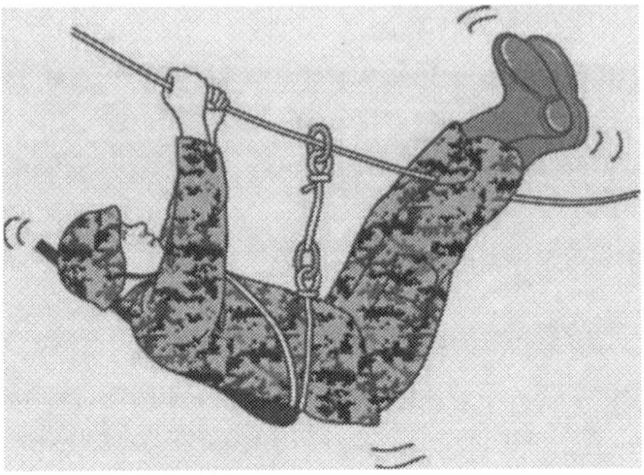

Swift Current Crossings

A single-rope bridge prevents being knocked down and swept away by a swift current. Use the following steps to move through a swift current:

Tie one end of a sling rope around your waist using a bowline.

Tie the running end of the sling rope in another bowline, and attach a carabiner to the bowline's loop.

Step up to the bridge. Face upstream.

Hook the carabiner to the single-rope bridge.

Walk sideways into the river while grasping the bridge rope in both hands.

Use the single-rope bridge for balance and remain standing, if possible.

Continue to move sideways through the river to the far shore.

Slow Current Crossings

If you face little or no current, it is not necessary to hook up to a bridge rope with a carabiner. Lie on your back in the water beneath the single-rope bridge. Support your body weight with your waterproof pack. Use the bridge rope and pull yourself hand over hand across the river.

Removal

If you are the last Marine waiting to cross, pull on the standing end of the rope to release the knot, then tie the rope around your waist using a bowline. The Marines on the far shore will pull you through the water.

Appendix
Knot Tying

Square Knot

The square knot is used to secure two ropes of equal diameter together so they form one continuous rope that will not slip.

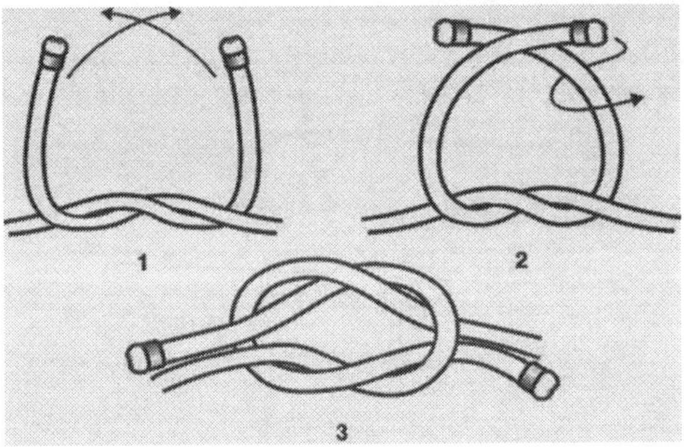

Bowline

The bowline forms a loop that will not tighten or slip under strain. It is easily untied.

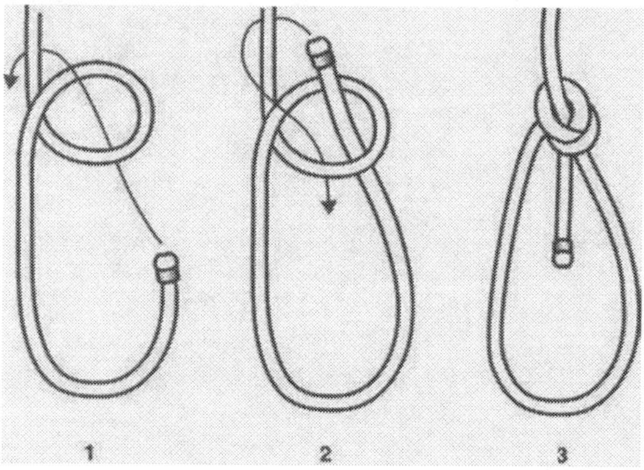

Hitches

Hitches are used to dress (prepare) knots and secure loose ends.

Half Hitch

The half hitch is used to tie a rope to a tree or to a larger rope. It will hold against a steady pull, but is not a secure hitch. It is frequently used to secure the free end of a rope.

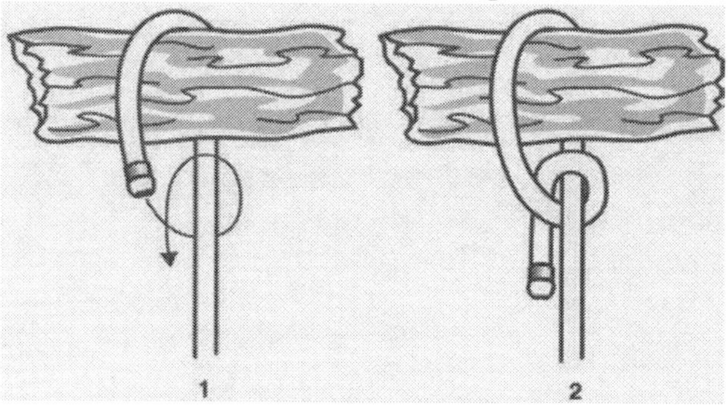

Two Half Hitches

Two half hitches can be used to secure the running end of a rope.

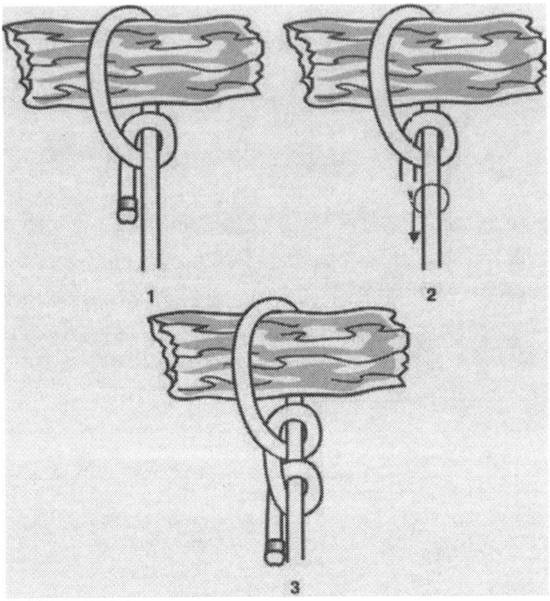

Round Turn and Two Half Hitches

A round turn and two half hitches can be used to fasten a rope to a tree. This hitch does not jam.

Butterfly Knot

The butterfly knot is used for anchor lines. Tied properly, it will not tighten on itself to the point that it cannot be easily untied. The butterfly knot can be used to pull a rope bridge taut. This knot can be used to tighten a fixed rope when mechanical means are not available. It will not jam if a stick is placed between the two upper loops.

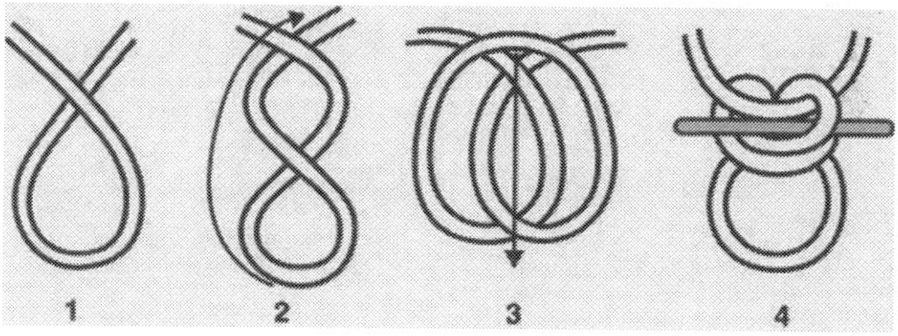

UNITED STATES MARINE CORPS
Marine Corps University

SCRS 1201
26 Mar 2001

USMC Land Navigation

Sergeants Course

LEARNING OBJECTIVES.

1. Terminal Learning Objective. With the aid of references and given an area of operations during day/night, military map, protractor, compass, and grid coordinates, navigate using basic techniques, per FM21-26. (SCRS 12.01)

2. Enabling Learning Objectives.

a. Without the aid of references and given a grid coordinate, state the distance in meters that the grid coordinate will locate a point within, per FM21-26 (SCRS 1201 01a)

b. Without the aid of references and given a military map, protractor, and ten (10) eight (8) digit grid coordinates, plot the points on the map, per FM21-26 (SCRS 1201.01b)

c. Without the aid of references and given a military map, protractor, and ten (10) eight (8) digit grid coordinates, measure the distance from one point to another, per FM21-26 (SCRS 1201.01c)

d. Without the aid of references and given a military map, protractor, and ten (10) eight (8) digit grid coordinates, determine the grid azimuth between two points, per FM21-26 (SCRS 1201.01d)

e. Without the aid of references state the procedures used to navigate using a lensatic compass, per FM21-26 (SCRS 1201.01e)

f. Without the aid of references and given a military map, protractor, and back azimuths from two (2) known points, plot a resection, in order to identify your position on the map, per FM21-26 (SCRS 1201.01f)

g. Without the aid of references, state the procedures used to by-pass obstacles, per FM21-26 (SCRS 1201.01g)

h. With the aid of references and given a checklist, area of operations, lensatic compass, military map of the area, eight (8) digit grid coordinates, starting point, finishing point, and as a member of a two (2) man team, navigate during daylight hours in order to locate five (5) checkpoints on a map of the area in a time limit of four (4) hours, per FM21-26. (SCRS 1201.01h)

i. With the aid of references and given a checklist, area of operations, lensatic compass, military map of the area, eight (8) digit grid coordinates, starting point, finishing point, and as a member of a two (2) man team, navigate during darkness hours in order to locate three (3) checkpoints on a map of the area in a time limit of four (4) hours, per FM21-26 (SCRS 1201.01i)

OUTLINE:

1. COMPONENTS OF A MAP. A map can be compared to any other piece of equipment. Before you use it, you must read the instructions. Map instructions are placed around the outer edge of the map and are called "marginal information." A topographical map is defined as a type of map which portrays terrain features, as well as the horizontal positions of the features represented. The vertical positions, or relief, are normally represented by contour lines. All maps are not the same, so every time you use a different map, examine the marginal information carefully.

a. Sheet Name. A map is named after the most prominent cultural or geographical feature. Whenever possible, the name of the largest city on the map is used. The sheet name is found in two places: the center of the upper margin and either the right or left side of the lower margin.

b. Sheet Number. The sheet number is used as a reference number for that map sheet. It is found in two places: the upper right margin and the lower left margin.

c. Series Name. The map series name is found in the upper left margin. It usually includes a group of similar maps at the same scale and/or the same sheet lines designed to cover a particular geographical area. The name given a series is that of the most prominent areas.

d. Scale. The scale note is a representative fraction that gives you the ratio of a distance on the map to the corresponding distance on the earth's surface. For example, the scale note 1:50,000 on many maps indicates that one inch on the map equals 50,000 inches on the ground. Maps with different scales will display different degrees of topographical detail. For example, a map with a scale of 1:25,000 will give more detail than a 1:50,000 map because one inch on the map represents only 25,000 inches on the ground, rather than the 50,000 inches of the 1:50,000 map. The scale is found both in the upper left margin after the series name and in the center of the lower margin.

e. Series Number. The series number is a sequence reference expressed either as a four-digit number (1125) or as a letter, followed by a three or four-digit number (M556; N3341). It is found in both the upper right margin and the lower left margin.

f. Edition Number. The edition number represents the age of the map in relation to other editions of the same map and the agency responsible for its production. For example, EDITION 5-DMATC indicates the fifth edition prepared by the Defense Mapping Agency Topographic Center. EDITION 6-DMA indicates the sixth edition prepared by the Defense Mapping Agency. Edition numbers run consecutively. A map bearing a higher edition number is assumed to contain more recent information than the same map bearing a lower edition number. It is found in the upper right margin and in the lower left margin.

g. Index To Boundaries. The index to boundaries diagram, which is a miniature of the map, shows the boundaries that occur within the map area, such as county lines and state boundaries. The index of boundaries diagram appears in the lower right margin of all map sheets.

h. Adjoining Sheets Diagram. Maps contain a diagram that illustrates the adjoining map sheets. The adjoining sheets diagram usually contains nine rectangles, but the number may vary depending on the locations of the adjoining sheets. All represented sheets are identified by their sheet numbers. The diagram is found in the lower right margin of all map sheets.

i. Elevation Guide. The elevation guide is a miniature characterization of the terrain shown. The terrain is shown by bands of elevation, spot elevations, and major drainage features. The elevation guide helps you rapidly identify major land forms. It is normally found in the lower right margin.

j. Declination Diagram. This indicates the angular relationships of true north, grid north, and magnetic north. Recent edition maps have a note indicating how to convert azimuths from grid to magnetic and from magnetic to grid next to the declination diagram. The declination diagram is located in the lower margin.

k. Bar Scales. Bar scales are used to convert map distance to ground distance. Maps may have three or more bar scales, each in a different unit of measure. Exercise care when using the scales, especially in the selection of the unit of measure. The bar scales are located in the center of the lower margin.

l. Contour Interval Note. The contour interval note states the vertical distance between adjacent contour lines on the map. When supplementary contours are used, the interval is indicated. In recent edition maps, the contour interval is given in meters instead of feet. This note is found in the center of the lower margin normally below the bar scales.

m. Legend. The legend illustrates and identifies the topographic symbols used to depict some of the more prominent features on the map, such as railroad tracks, buildings, and swamps. The symbols are not always the same on every map. Always refer to the legend to avoid error when reading a map. The legend is located in the lower left margin.

2. LOCATE GRID COORDINATES ON A MAP.

a. Grid lines. Grid lines are a series of straight lines intersected at right angles and forming a series of squares. It furnishes the map reader with a system of squares similar to the block system of most city streets. Two digits are printed in large type at each end of the grid lines, and these same two digits appear at intervals along the grid lines on the face of the map. They are called principal digits. They are of major importance to the map reader because they are the numbers he will use most often for referencing points.

b. Vertical Grid Lines. These run from the bottom of the map sheet to the top of the sheet (grid south---grid north). They function as the left and right (eastern and western) boundaries of a grid square and are labeled in the margin.

c. Horizontal Grid Lines. These run from the left side of the map sheet to the right side of the map sheet (grid west---grid east). They function as the bottom and top boundaries (southern and northern) of a grid square and are labeled in the margin.

d. Grid Squares. These intersect at right angles of the horizontal and vertical grid lines. The most common military map contains grid squares that measure 1000 meters by 1000 meters (not 1000 square meters as many people think). Any point located within the grid square is considered to be part of the grid square.

e. Basic Map Reading Rule. The designation of a point is based on the principle: Read right then up. Always read right on the vertical grid lines then up on the horizontal grid lines.

f. Grid Square Identification. It is important that all of you understand how to apply the map reading rule to identify a grid square and locate a point within a grid square.

(1) The coordinates of a grid square are found by combining the values of the vertical and horizontal grid lines that form the lower left-handed corner of that grid square.

(2) First read right to the vertical grid line that forms the left (western) boundary of the grid square and record the principal digits.

(3) Next read up on the horizontal grid line that forms the left (western) boundary of the grid square and record the principal digits.

(4) The combination of the principal digits that label the vertical grid line and horizontal grid line are the identification of the grid square or its coordinates.

(5) A four digit grid coordinate locates a point to within 1000 square meters, on the map, which is called a grid square.

(6) A six digit grid coordinate will locate a point on a map within 100 meters.

(7) An eight digit grid coordinate will locate a point on a map within 10 meters

3. PLOTTING POINTS ON A MAP.

a. 6-Digit Grid Coordinates. Imagine dividing a grid square into 100 smaller squares. The coordinates of a point in such a grid square will have six digits (numbers). Each of the grid squares in figure 1201-1 is 1,000 meters long and 1,000 meters high. One of the grid squares is divided into 100 smaller squares--each 100 meters long and 100 meters high. Note that the lines within the grid square also read RIGHT, then UP. In a 6-digit grid coordinate, such as 284936, the first 3 numbers are the "read right" part of "read right, then up;" whereas, the last 3 numbers are the "then up" part. To locate point C in grid square 3050 (see figure 1201-1), you should use the following procedure:

(1) Read right to vertical line 30. Note that point C is 3/10 of the distance toward the vertical line 31. Write the vertical coordinate for point C as 303.

(2) Read up vertical line 30 to horizontal line 50. Note that point C is 7/10 of the distance toward the horizontal line 51. Write the horizontal coordinate of point C as 507.

(3) To write the 6-digit coordinate that locates point C, simply combine the two readings--303507. The vertical reading is ALWAYS placed before the horizontal reading.

NOTE: Do not actually draw the lines within the grid square (see figure 1201-1), since such lines would obstruct other information. You can determine a 6-digit grid coordinate by approximation, or by using your coordinate scale on a protractor. Use of the coordinate scale is explained in the next paragraph.

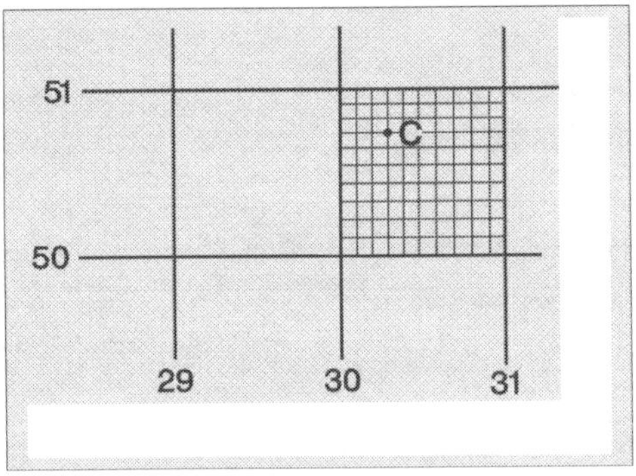

Figure 1201-1.

b. Protractor. If you require a more accurate 6-digit grid coordinate than can be obtained by estimation, you should use a coordinate scale. There are three coordinate scales located on your protractor: 1:100,000, 1:50,000, and 1:25,000. Use the one that corresponds with the scale of the map you are using. In most cases this will be 1:50,000. To locate point C in grid square 3050 using the coordinate scale on your protractor, use the following procedures:

(1) Place the proper coordinate scale of your protractor on the map so that the zero-zero point is to the bottom right of the scale (see figure 1201-2).

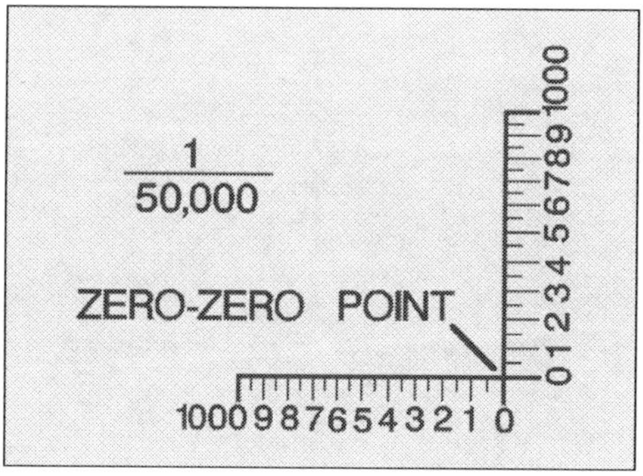

Figure 1201-2

(2) Place the zero-zero point at the lower left hand corner of grid square 3050 (see figure 1201-3).

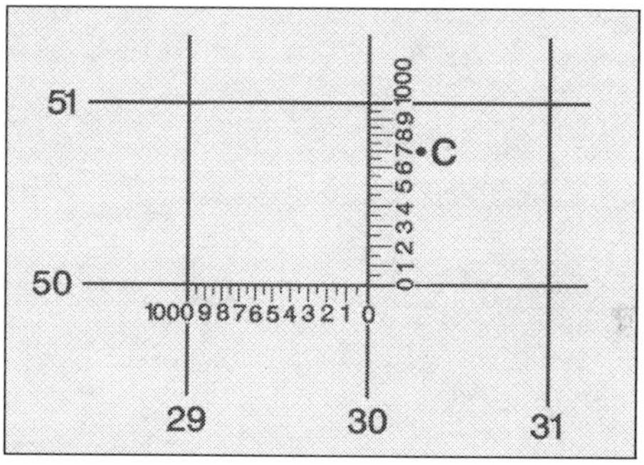

Figure 1201-3

(3) Keeping the horizontal line of the scale directly on top of the east-west grid line (in this case, grid line 50), slide it to the right until the vertical line of the scale touches the point (point C) for which the coordinates are desired (see figure 1201-4)

(4) Examine the two sides of the coordinate scale to ensure that the horizontal line of the scale is aligned with the east-west grid line, and the vertical line of the scale is parallel with the north-south grid line.

(5) Determine your RIGHT reading by first reading the value of the grid line to the left of point C (30). Add to this value the number that tells how far into the grid square point C is. In this case, it is 300 meters (see figure 1204-4). You now have the complete right reading of 303.

(6) Next , determine your UP reading by first reading the value of the horizontal grid line below point C (50). Add to this value the number which tells how far point C is up in the grid square. In this case it is 700 meters. You now have the complete UP reading of 507 (see figure 1201-4). When determining both your right and up reading, round your value to the closest number on your coordinate scale.

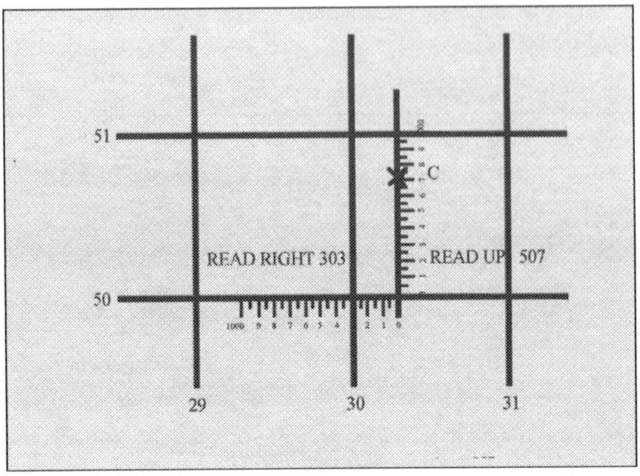

Figure 1201-4.

(7) By combining the RIGHT reading (303) with the UP reading (507), you have accurately determined that the 6-digit grid coordinate of point C is 303507.

c. Locating Points On A Map. You have learned how to determine the grid coordinates of a given point on a map to within 100 meters (6-digits). Next you need to know the method of locating a point on your map when given a 6-digit grid coordinate. To locate a point on the map, using the coordinate scale on your protractor, use the following procedures:

(1) Locate The Correct Grid Square. To locate the correct grid square, determine the 4-digit grid coordinates from the given 6-digit grid coordinates.

(a) Split your 6-digit coordinates into two parts. For example, the grid coordinates 025672 would be split to read 025 for your RIGHT reading, and 672 for your UP reading.

(b) Determine the vertical (north-south) grid line. It is the first two numbers of your RIGHT reading.

02

(c) Determine the horizontal (east-west) grid line. It is the first two numbers of your **UP** reading.

67

(d) Determine your 4-digit grid coordinates. Combine the **RIGHT** with the **UP** reading.

02 + 67 = 0267

(2) Plot The 6-Digit Coordinates. The procedures for plotting your grid coordinates are listed as follows:

(a) Place the proper coordinate scale of your protractor with the zero-zero point at the lower left hand corner of the grid square 0267, keeping the horizontal line of the scale directly on top of the east-west grid line.

(b) Examine the two sides of the coordinate scale to ensure that the horizontal line of the scale is aligned with the east-west grid line, and the vertical line of the scale is parallel with the north-south grid line (see figure 1205-5).

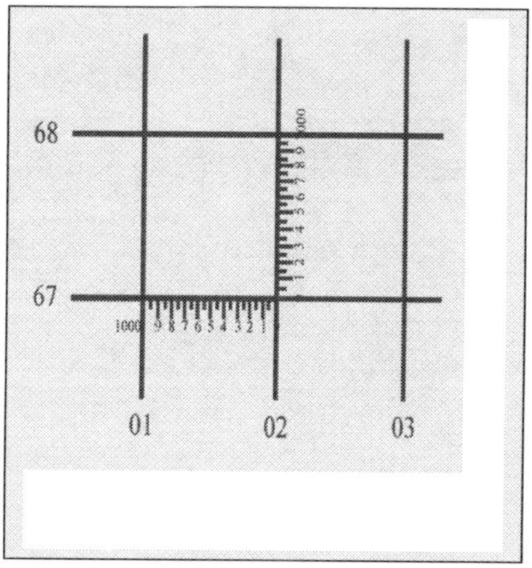

Figure 1201-5.

(c) Slide the protractor to the right until vertical grid line 02 intersects the horizontal scale at the 100-meter reading "5" (see figure 1201-6). This point 025, is your RIGHT reading.

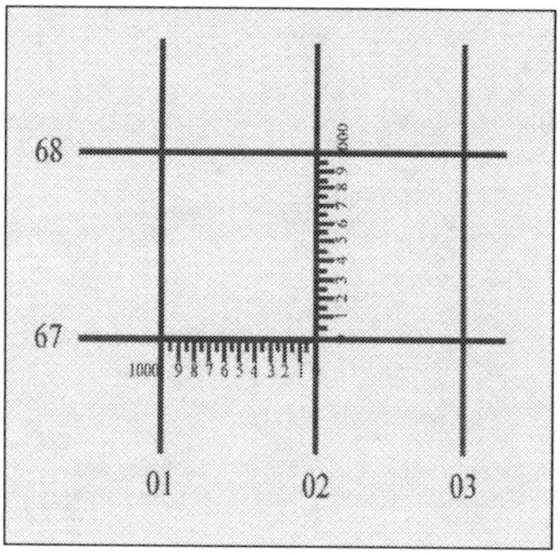

Figure 1201-6.

(d) Determine the position of your UP reading by plotting a point adjacent to your vertical scale equal to your UP reading. In this case the UP value of the vertical scale is at the 100-meter reading 2 (see figure 1201-7). This point is not only your UP reading, 672, but is also the location of your 6-digit coordinate, 025672.

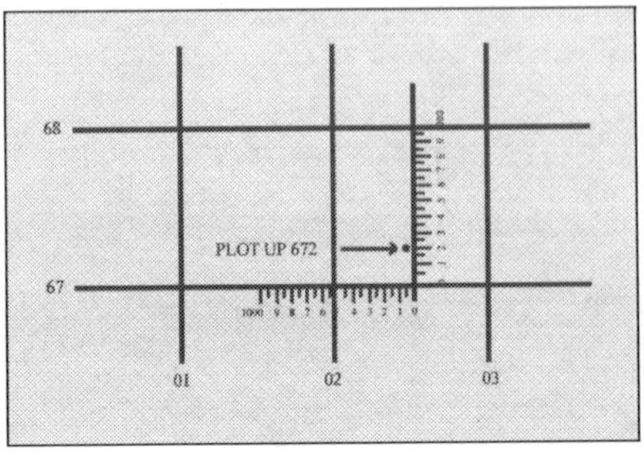

Figure 1201-7

d. <u>8-Digit Grid Coordinates</u>. In some mapping situations, it is desirable to further divide a 100-meter grid square into 10-meter grid squares. This is done in the same manner as dividing a 1,000 meter grid square into 100-meter grid squares, either through estimation or by use of a coordinate scale. The result is an 8-digit grid coordinate which identifies a point on the map to within 10 meters. However, 6-digit grid coordinates, when accurately determined, should meet all your needs for effective land navigation.

4. <u>CONTOUR LINES</u>. A contour line is a line representing an imaginary line on the ground along which all points are the same elevation.

a. Contour lines have certain things in common.

(1) Contour lines indicate a vertical distance above or below the datum plane. (Example datum plane - sea level.)

(2) The vertical distance between adjacent contour lines is known as contour interval, and the amount of the contour interval is given in marginal information.

(3) Starting at sea level, the zero contour, each contour line represents an elevation above sea level.

(4) On most maps the contour lines indicate the nature of the slope.

b. <u>Types Of Contour Lines</u>.

(1) <u>Index Contour Line</u>. Starting at zero elevation, every fifth contour line is drawn with a heavier line. These are known as index contours. Someplace along each index contour the line is broken and its elevation is given.

(2) <u>Intermediate Contour Lines</u>. The contour lines falling between index contours are called intermediate contour lines. They are drawn with a finer line than the index contours and do not have their elevation given.

c. <u>Determining Elevation</u>. Using the contour lines on a map, the elevation of any point may be determined by:

(1) Finding the contour lines on a map from the marginal information, and noting both the amount and the unit of measure.

(2) Finding the numbered contour line nearest the point for which the elevation is being sought.

(3) Determining the direction of the slope from the numbered contour lines (index contour) to the desired point.

(4) Counting the number of contour lines that must be crossed to go from the numbered index contour line to the desired point and noting the direction - up or down. The number of lines crossed, multiplied by the contour interval is the distance above or below the starting value.

(a) If the desired point is on a contour line, its elevation is that of the contour line.

(b) For a point between contours most military needs are satisfied by estimating the elevation to an accuracy of one half the contour interval.

(c) To estimate the elevation to the top of an unmarked hill, add half the contour interval to the elevation of the higher contour line around the hill.

(d) To estimate the elevation of the bottom at a depression, subtract half the contour interval from the value of the lowest contour line around the depression.

5. TYPES OF TERRAIN FEATURES. As you plan and execute land navigation, you must be able to recognize and associate ground forms which you see or expect to see on the ground with the same features shown by contour lines on the map. Once you can recognize these ground forms and understand their characteristics, you can better use certain terrain features to your advantage and avoid those which might hinder you.

a. Hill.

(1) Representation. A hill is an area of high ground. From a hilltop, the ground slopes down in all directions. A hill is shown on a map by contour lines forming concentric circles. The inside of the smallest closed circle is the hilltop.

(2) Application To Navigation. Your ability to recognize hills on a map will help you greatly in land navigation. You can plan your routes to avoid unnecessary travel over them. Checkpoints can be chosen on or near a prominent hilltop. You can use a distant prominent hilltop to guide you. Identifying hilltops on the ground and on the map can help you locate your position accurately (see figure 1201-8).

Figure 1201-8

b. Ridge.

(1) Representation. A ridge is a series of hills that are connected to each other near the top. A ridge line may extend for many miles. It may be winding or quite straight. It may have a reasonably uniform elevation along its top or it may vary greatly in elevation.

(2) Application To Navigation. A ridge often serves as an ideal reference line during both day and night. If a map study shows that the ridge is constantly on your left, then as long as you can see the familiar characteristics of the ridge silhouetted against the skyline on your left, you know that you are heading in the proper general direction. After passing through thick terrain, if the ridge line is no longer on your left, you know you are heading in the wrong general direction. It is normally easier to maintain direction, observe steering marks, and move with less obstruction by moving along the top of the ridge as opposed to the sides of the ridge (if the tactical disadvantage of being silhouetted against the skyline is not a determining factor, and if the top of the ridge is fairly uniform) (see figure 1201-9).

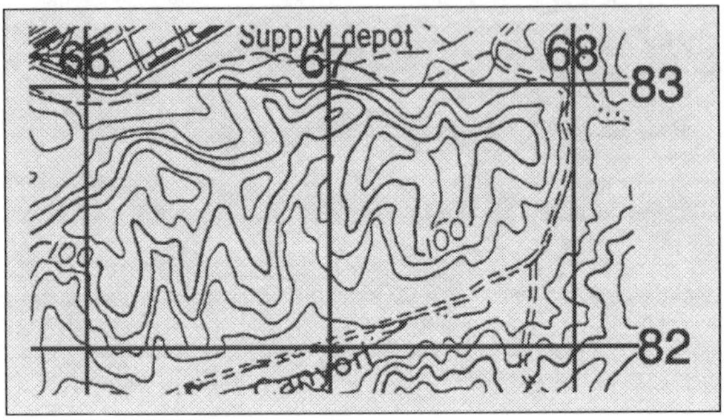

Figure 1201-9.

c. Saddle.

(1) Representation. This is a dip or low point between two areas of higher ground. A saddle is not necessarily the lower ground between two hilltops; it may be simply a dip or break along a level ridge crest. If you are in a saddle, there is high ground in two opposite directions and low ground in the other two directions. A saddle is normally represented as an hourglass or by figure-eight shaped contour lines.

(2) Application To Navigation. A distinct, unmistakable saddle will often provide a suitable checkpoint or steering mark. Also, if you must cross a ridge, it is normally easiest to cross it at a saddle (see figure 1201-10).

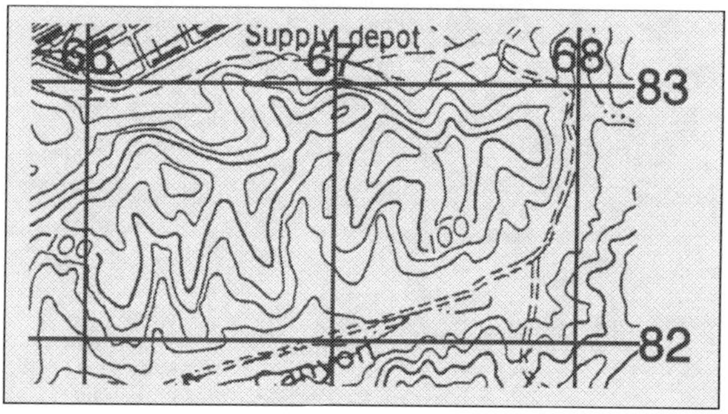

Figure 1201-10.

d. Finger.

(1) Representation. A finger is a short, continuous sloping line of higher ground, normally jutting out from the side of a ridge or hill. A finger is often formed by two roughly parallel draws. The ground slopes down in three directions and up in one. Contour lines on a map depict a finger with the U or V pointing away from high ground.

(2) Application To Navigation. A distinct, unmistakable finger may be chosen as a checkpoint or steering mark. In the defense, fingers provide good fields of view for observation posts. During movement, there is a chance of being skylined when traveling on fingers (see figure 1201-11).

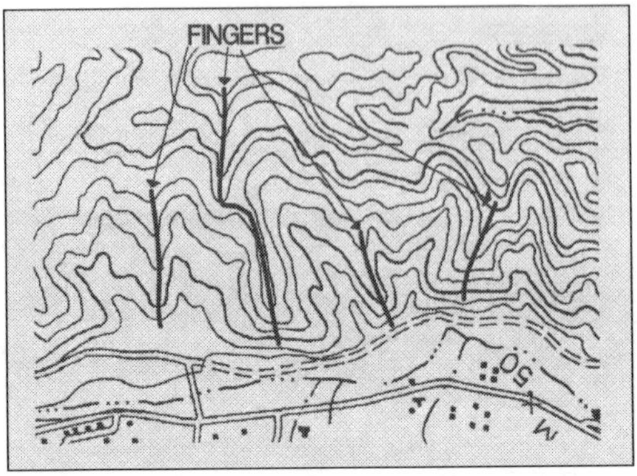

Figure 1201-11

e. Draw.

(1) Representation. A draw is a short, continuous sloping line of low ground, normally cut into the side of a ridge or hill. Often, there is a small stream running down the draw. In a draw, there is essentially no level ground. Therefore, little or no maneuver room exists within its confines. If you are standing in the middle of a draw, the ground slopes upward in three directions and downward in the other direction. Contour lines (see figure 1201-12) on a map depict a draw with the U or V pointing toward high ground.

(2) Application To Navigation. A draw may or may not be a wise route to follow. The sides of a draw are often steep. The sides of draws are normally steeper than the fingers on both sides of the draw. Fingers normally have convex slopes while draws normally have concave slopes. Therefore, when you must climb a hill, draws provide an easy route at the bottom of the hill, but the route gets more difficult near the top when the sides of draws are normally quite steep. Draws often provide good concealment during movement, but if the draw is covered by an enemy machine gun, it could become a death trap due to its steep sides. A draw can be used to maintain direction.

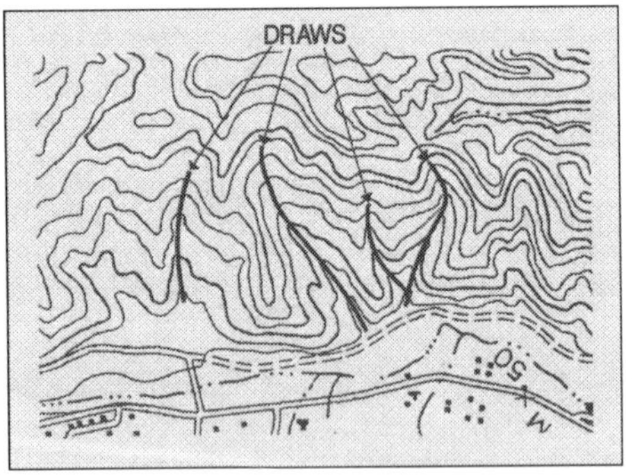

Figure 1201-12

(3) Distinguishing Between Draws And Fingers. Look at point "A" on figure 1201-13. Is it a finger or a draw? If you think it is a finger, you are wrong. If you think it is a draw, you also are wrong. You see, from the

amount of information given in the figure, you can not tell what type of terrain feature it is. You MUST look at the surrounding terrain features. If there is a hill in the direction of the open end of the "U" then the feature is a finger; if not, then it is a draw. If there is a stream in the middle of the feature, then is it a finger or a draw? That's right, a draw. Along those lines, here's a trivia question for you - which way does the closed end of the "U" of the contour lines point on a stream, upstream or downstream? You already have enough information to answer this. Which way does the closed end of the "U" of the contour lines point in a draw? And where are streams, in draws or fingers? So the correct answer is that the closed end of the "U" points UPSTREAM.

Figure 1201-13

f. Valley.

(1) Representation. A valley is continuous line of low ground that normally rests in between two roughly parallel ridges. Valleys often have streams or rivers running through them.

(2) Application To Navigation. If you are navigating through a valley that has a stream, keep in mind that while a dry stream bed provides a relatively vegetation-free, easy-to-follow route, it will often meander back and forth across the valley floor. During inclement weather, many streams are subject to flash flooding. Additionally, the banks of streams, creeks, and rivers often contain mud flats, marshes, and extremely thick vegetation. During tactical foot movements, keep in mind the need for security and silent movement. Moving silently while wading in water requires a great deal of effort on the part of each Marine. Therefore, while the trace of a stream will often provide an ideal reference line, it is recommended that you move away from the stream and use the valley as a reference point.

g. Cuts And Fills.

(1) Representation.

(a) A CUT is a man-made feature resulting from cutting through high ground, usually to form a level bed for a road or railroad track. Cuts are drawn with a contour line along the cut line. This contour line extends the length of the cut and has tick marks that extend from the cut line to the roadbed, if the map scale permits this level of detail (see figure 1201-14).

(b) A FILL is a man-made feature resulting from filling a low area, usually to form a level bed for a road or railroad track. Fills are drawn with contour lines along the fill line. This contour line extends the length of the filled area and has tick marks that point toward lower ground. If the map scale permits, the length of the fill tick marks are drawn to scale and extend from the base line of the fill (see figure 1201-14).

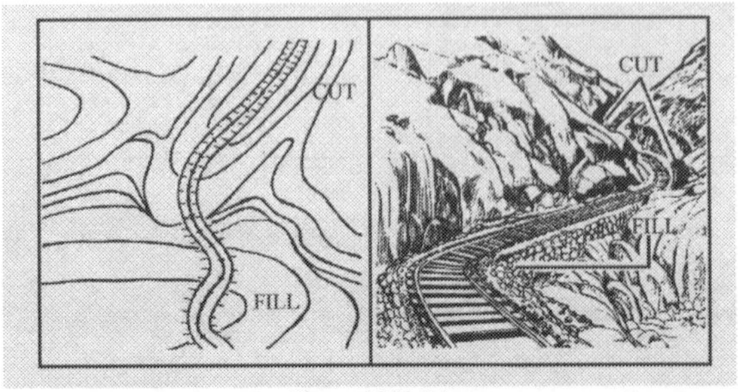

Figure 1201-14

(2) Application To Navigation. Prominent cuts and fills can often be used as checkpoints. If the map shows that a road or railroad is punctuated with cuts and fills, it indicates that efforts have been made to eliminate radical changes in slope.

h. Depression.

(1) Representation. This is a low point in the ground or a sinkhole. It is an area of low ground surrounded by higher ground in all directions, or simply a hole in the ground. Usually only depressions that are equal to or greater than the contour interval will be shown. On maps, depressions are represented by closed contour lines that have tick marks pointing toward low ground (see figure 1201-15).

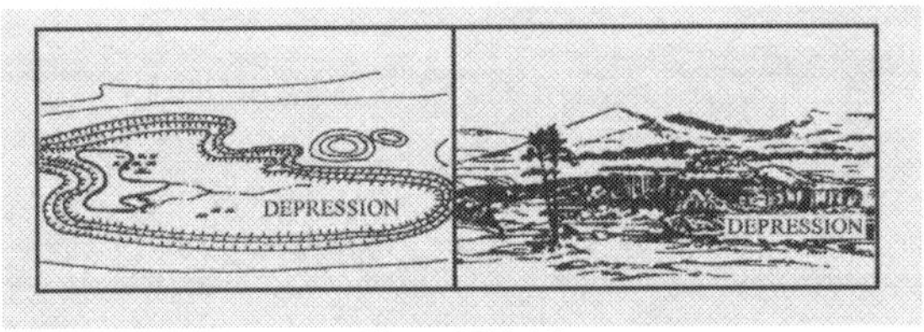

Figure 1201-15

(2) Application To Navigation. Prominent depressions can often be used as checkpoints. Because of the usually steep sides of depressions, bypass them if possible.

i. Cliff.

(1) Representation. A cliff is a vertical or near vertical terrain feature. It is an abrupt change of the land. When a slope is so steep that the contour lines converge into one "carrying" contour of contours, this last contour line sometimes has tick marks pointing toward low ground (see figure 1201-16).

(2) Application To Navigation: Prominent cliffs can often be used as checkpoints. Because of the steep face of a cliff, it should be bypassed.

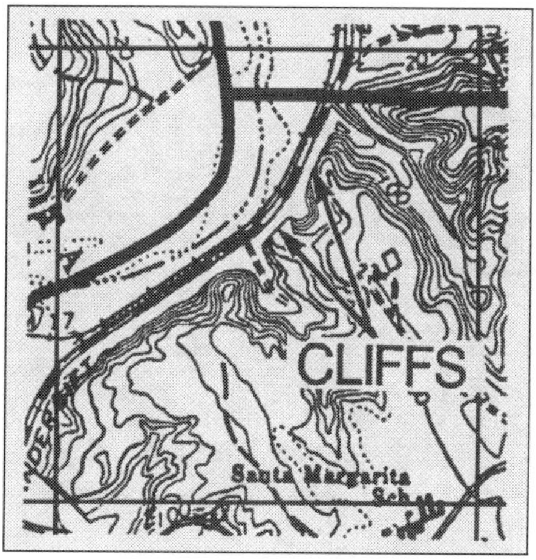

Figure 1201-16

6. DETERMINE DISTANCE ON A MAP.

a. Determining Straight-Line Distance.

(1) To determine straight-line distance between two points on a map, lay a straight-edged piece of paper on the map so that the edge of paper touches both points and extends past them. Make a tick mark on the edge of the paper at each point (see figure 1201-17). Remember that the center of the topographic symbol accurately designates the true location of the object on the ground; therefore, measure all map distances from the center of the topographic symbol.

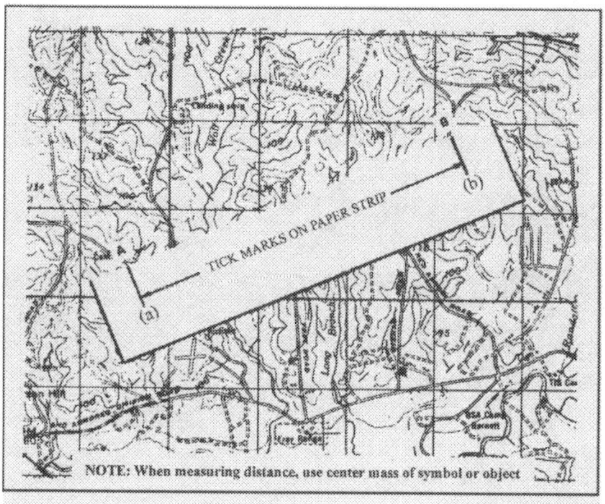

Figure 1201-17

(2) To convert map distance to ground distance, move the paper down to the appropriate unit of measure on the graphic bar scale, and align the right tick mark (b) with a printed number in the primary scale so that the left tick mark (a) is in the extension scale (figure 1201-18).

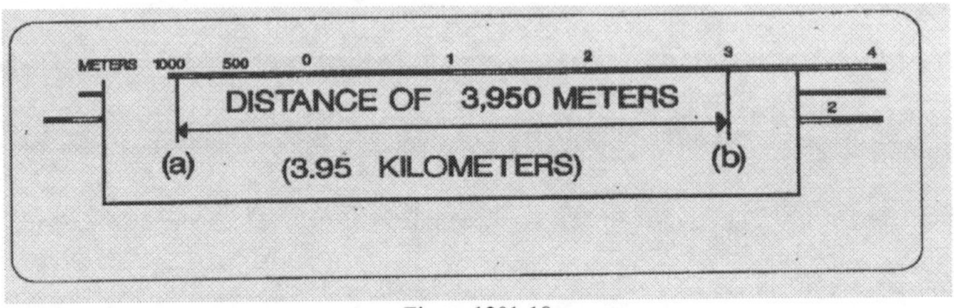

Figure 1201-18

In this case the right tick mark (b) is aligned with the 3,000 meter mark in the primary scale, thus the distance is at least 3,000 meters. To determine the distance between the two points to the nearest 10 meters, look at the extension scale. The extension scale is numbered with zero at the right and increases to the left. When using the extension scale, always read RIGHT TO LEFT (see figure 1201-18). From the zero to the end of the first shaded square is 100 meters. From the beginning of the white square to the left is 100 to 200 meters; at the beginning of the second shaded square is 200 to 300 meters. Remember, the distance in the extension scale increases from right to left. To determine the distance from tick mark (a), estimate the distance inside the squares to the closest tenth. As you break down the distance between the squares in the extension scale, you will see that tick mark (a) is aligned with the 950-meter mark. Adding the distance of 3,000 meters determined in the primary scale, we find that the total distance between (a) and (b) is 3,000 + 950 = 3,950 meters.

(3) There may be times when the distance you measure on the edge of the paper exceeds the graphic scale. One technique you can use to determine the distance is to align the right tick mark (b) with a printed number in the primary scale, in this case 5 kilometers (see figure 1201-19). You can see that from point (a) to (b) is more than 6,000 meters. To determine the distance to the nearest 10 meters, place a tick mark (c) on the edge of the paper at the end of the extension scale (see figure 1201-19).

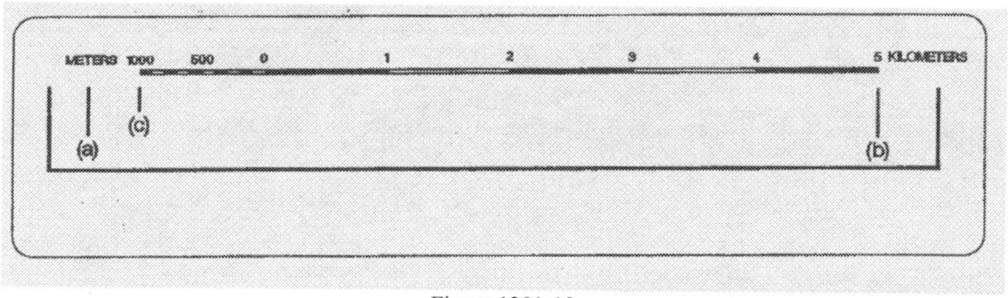

Figure 1201-19

You know that from point (b) to (c) is 6,000 meters (5,000 from the primary scale and 1,000 from the extension scale). Now, measure the distance between points (a) and (c) on your sheet of paper the same way you did earlier, only use point (c) as your right hand tick mark (see figure 1201-20). The total ground distance between start and finish points is 6,420 meters.

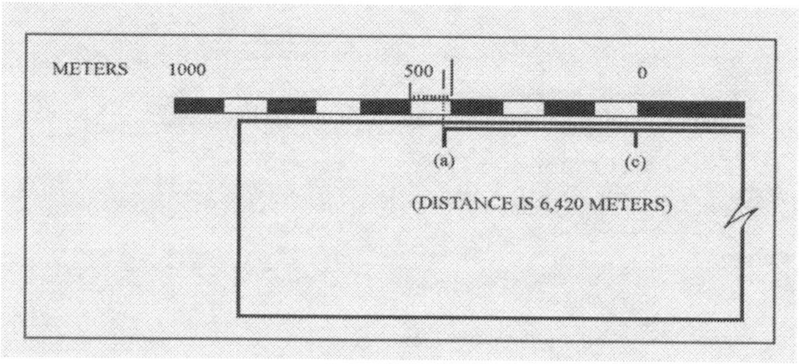

Figure 1201-20

(4) One point to remember is that distance measured on a map does not take into consideration the rise and fall of the land. All distances measured by using the map and graphic scales are flat distances. Therefore, the distance measured on a map will increase when actually measured out on the ground (see figure 1201-21).

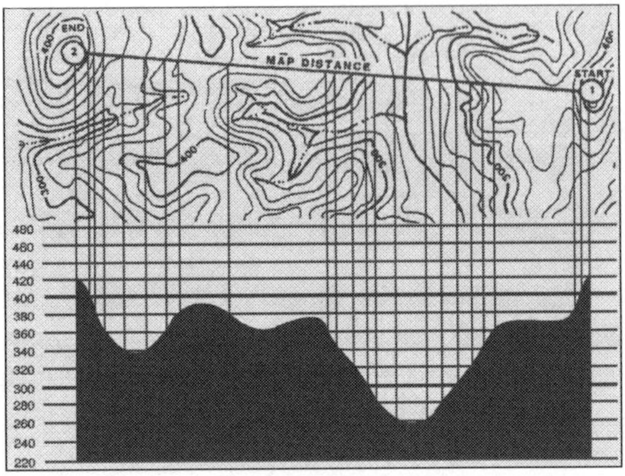

Figure 1201-21.

b. Determining Irregular Map Distance. To measure distance along a winding road, stream, or other curved line, you still use the straight edge of a piece of paper. In order to avoid confusion concerning the starting point and the ending point, a six-digit coordinate, combined with a description of the topographical feature, should be given for both the starting and ending points. Place a tick mark on the paper and map at the beginning point from which the curved line is to be measured. Place a paper strip or other material with a straightedge along the center of the irregular feature (see figure 1201-22), and extend the tick mark onto the paper strip. Because the paper strip is straight and the irregular feature is curved, the straightedge will eventually leave the center of the irregular feature. At the exact point where this occurs, place a tick mark on both the map and paper strip.

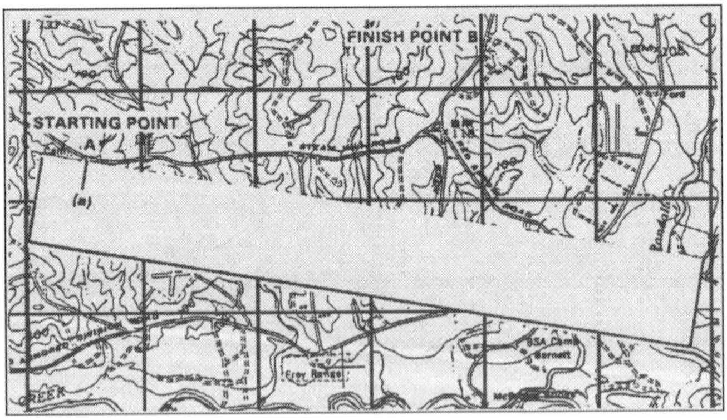

Figure 1201-22

Keeping both tick marks together (on paper and map), place the point of the pencil close to the edge of the paper on the tick mark to hold it in place and pivot the paper until another straight portion of the curved line is aligned with the edge of the paper. Repeat this procedure while carefully aligning the straightedge with the center of the feature and placing tick marks on both the map and paper strip each time it leaves the center until you have ticked off the desired distance (see figure 1201-23).

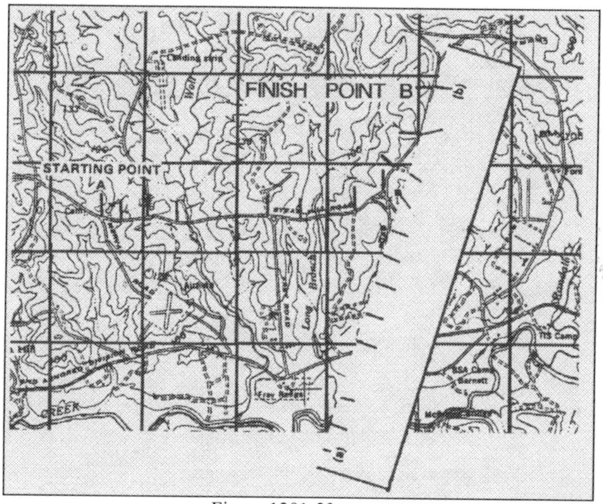

Figure 1201-23.

Place the paper strip on a graphic bar scale and determine the ground distance measured.

7. PLOTTING A GRID AZIMUTH.

a. Azimuth Defined. An azimuth is a straight line from one point to another. This definition, however, leaves much to be desired in understanding the complete meaning of the word azimuth as it applies to land navigation. Therefore, the following definition is required: An azimuth is an angle measured in a clockwise direction from a predetermined base line. Before attempting to determine or follow an azimuth in the field, you must have a clear understanding of each part of this definition.

(1) An Azimuth Is An Angle. "An azimuth is an angle," means that it is part of a circle. But just how much of a circle does each azimuth represent? If a circle were divided into 360 equal "slices of pie" each slice would be one degree (see figure 1201-24).

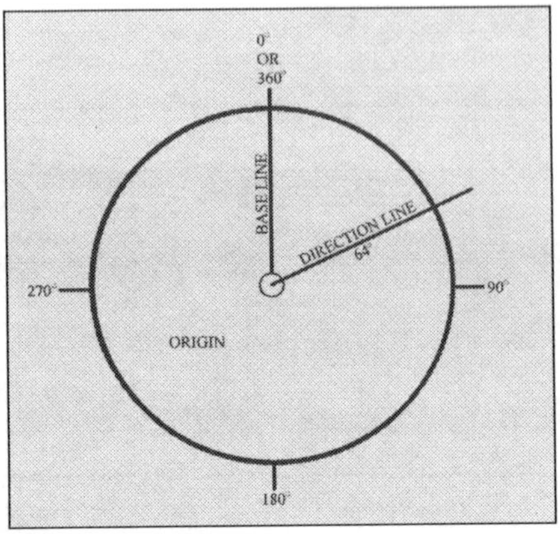

Figure 1201-24

(2) Measured In A Clockwise Direction. When we say that an azimuth is an angle measured in a clockwise direction, we mean that each of the angles discussed above must have a starting point and from there progress in numerical value in a clockwise direction around the circle until they return to the starting point. The starting point has a value of 0 degrees, or (since it is also the final direction line) 360 degrees. It may be expressed as either 0 degrees or 360 degrees. Since the degree value of azimuths always progresses in a clockwise direction, all azimuths between 0 degrees and 180 degrees will be on the right side of the imaginary circle and all azimuths between 180 degrees and 360 degrees will be on the left side of the circle (see figure 1201-24). Keep in mind that there are only 360 degrees in a circle. When working map problems, if you mathematically arrive at a figure exceeding 360 degrees, then you have gone completely around the circle and started over again. For example, if you add 15° to 350°, it would be expressed as 5°, NOT 365°.

(3) From A Predetermined Base Line. When we say that an azimuth is an angle measured in a clockwise direction from a predetermined base line, we mean that it is a certain number of degrees measured in a clockwise direction from some sort of reference point. It is this portion of the definition that causes the most misunderstanding, confusion, and often loss of direction in the field. This reference point or base line we are referring to is north. There are three base lines--true north, magnetic north, and grid north (see figure 1201-25). The most commonly used are magnetic and grid north.

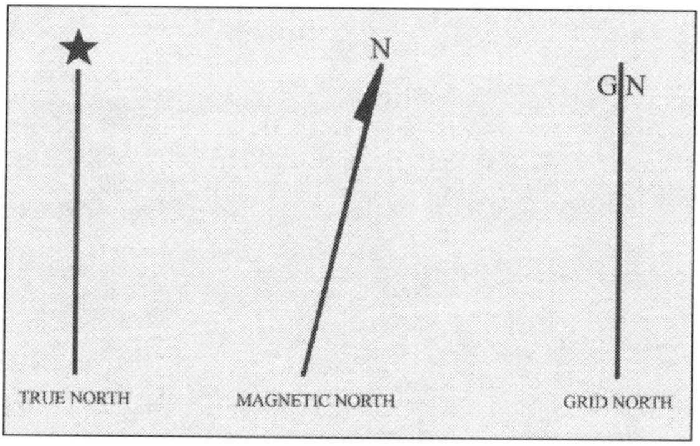

Figure 1201-25

(a) True North. The true north line is a line from any point on the earth's surface to the North Pole. True north can be found at night by locating the North Star, which always points towards true north. True north is usually represented on the declination diagram by a line ending with a star (see figure 1201-25). True north is used almost exclusively when navigating without a compass.

(b) Magnetic North. The earth has a magnetic field that is close to (but not exactly on) the North Pole. The direction to this north magnetic pole is indicated by the north-seeking arrow of your lensatic compass. Magnetic north is usually symbolized on the declination diagram by a line ending with a half arrowhead (see figure 1201-25). Anytime you use the compass to plan or follow an azimuth in the field, you must work with azimuths measured from magnetic north.

(c) Grid North. This base line is established by using the vertical grid lines on the map. Grid north may be symbolized on the declination diagram by the letters GN (see figure 1201-25). Anytime you use a protractor in conjunction with a vertical grid line to determine or plot an azimuth on a map, you must work with an azimuth measured from grid north.

(4) When using grid north and magnetic north as base lines, you need to understand the difference between a grid azimuth and a magnetic azimuth:

(a) A grid azimuth is an angle measured in a clockwise direction from grid north.

(b) A magnetic azimuth is an angle measured in a clockwise direction from magnetic north.

b. Determining A Grid Azimuth. There are two methods of measuring grid azimuths from one point to another on the map. Whichever method you use, remember that you are dealing with GRID AZIMUTHS. A grid azimuth CAN NOT be followed with a compass.

(1) Protractor And String Method. To use this method, you must first modify your protractor. Using a needle and piece of thread, punch a SMALL hole through the index mark of your protractor. (The thread should be about 6 inches long.) Tie a knot in the thread on each side of the protractor as close to the index line as possible to secure the thread to the protractor. Now you are ready to go.

(a) Place the index mark on your starting point. Ensure that the vertical base line is parallel with a north-south grid line and the horizontal base line is parallel with an east-west grid line.

(b) Holding the protractor firmly against the map with one hand, stretch the piece of thread with your other hand so that the thread intersects your second point.

(c) The point where the thread intersects the inside scale of the protractor is your azimuth (the outside scale is in mils). This is the GRID AZIMUTH from the starting point to the second point.

(2) Protractor and Pencil Method. Before we get started with the steps for this method, you need to understand something. If you go in a STRAIGHT line from one point to another, does the azimuth ever change? Of course not. No matter where you are on that line of march, as long as you maintain a straight line, the azimuth never changes. Remember that when you get to Steps 1 and 3 for this method.

(a) Using a straight edge, draw a line connecting the two points. This line needs to be at least 4 inches long so that the line reaches the edge of the protractor.

(b) Label the two points A and B, with your point A being your starting position.

(c) Place the index mark of your protractor on the line you just drew where that line intersects a vertical grid line. This point should be as close to point A as possible. Ensure that the entire vertical base line on your protractor is directly on the vertical grid line that your line intersects. If it is not, then your azimuth reading will be inaccurate.

NOTE: The reason for placing the index mark on a point where your line intersects a vertical (or horizontal) grid line instead of directly on point A is so that you get a more accurate reading. Remember, as long as you are on that straight line from point A to point B, the azimuth does not change.

(d) The point where the line intersects the inside scale of the protractor is your azimuth (the outside scale is in mils). This is the GRID AZIMUTH from point A to point B.

c. Plotting A Grid Azimuth. To plot a grid azimuth on a map, follow these steps:

(1) Place the protractor on the map with the index mark at center mass of the known point.

(2) Ensure that the protractor's vertical base line is parallel with the closest north-south grid line and the horizontal base line is parallel with an east-west grid line.

(3) Make a mark on the map at the desired grid azimuth.

(4) Remove the protractor and draw a line connecting the known point and the mark on the map. You have now plotted the grid azimuth.

(5) To plot a known distance onto that azimuth, use a piece of paper to transfer the distance from the scale to the plotted azimuth. Remember, the azimuth you have just plotted is a GRID AZIMUTH. It can not be followed using a compass.

d. Determining A Back Azimuth. A back azimuth is the opposite direction of an azimuth. It is the same as doing an "about face." To obtain a back azimuth from an azimuth, ADD 180 degrees if the azimuth is 180 degrees or less or subtract 180 degrees if the azimuth is 180 degrees or more. The back azimuth of 180 degrees may be stated as 0 degrees or 360 degrees. So, what happens if you add when you should have subtracted or subtracted when you should have added? You will come up with either a negative number, or one that is over 360 degrees. Since you cannot express azimuths as either negative numbers or as numbers over 360 degrees then simply do the opposite of what you did before. If you added 180 degrees, then subtract. If you subtracted 180 degrees, then add.

8. NAVIGATE USING THE LENSATIC COMPASS.

a. Lensatic Compass Description. The lensatic compass consists of three major parts:

- cover
- base
- rear sight

(1) Cover. This protects the floating dial and the glass encasement. It contains the sighting wire and two luminous sighting dots for night navigation.

(2) Base.

(a) Floating Dial. This is mounted on a pivot so that it rotates freely when the compass is held level. It contains the magnetic needle. A luminous arrow and the letters "E" and "W" are printed on the dial. The arrow points to magnetic north. Letters fall at the east (E) 90 degrees and (W) 270 degrees. There are two scales: outer - denotes MILS (black); inner - denotes DEGREES (red).

NOTE: Mil, is another unit of measure. The mil (abbreviated m), is mainly used in artillery, tank, and mortar gunnery. The mil expresses the size of an angle formed when a circle is divided into 6,400 angles with the vertex of the angles at the center of the circle. A relationship can be established between degrees and mils. A circle equals 6400 mil divided by 360 degrees, or 17.78 mils. To convert degrees to mils, multiply degrees by 17.78.

(a) Glass Encasement. This houses the floating dial and contains a fixed black index line.

(b) Bezel Ring. This device that clicks when turned. It contains 120 clicks when rotated fully. Each click equals 3 degrees. A short luminous line is used in conjunction with the north-seeking arrow during night navigation.

(c) Thumb Loop. This is attached to the base.

(3) Rear Sight. This is used to lock the floating dial. The rear sight must be opened more than 45 degrees to allow the floating dial to float freely.

(a) Lens. This is used to read the floating dial.

(b) Rear Sight Slot: This is used in conjunction with the front sighting wire when aiming at objects.

b. Handling The Compass.

(1) Inspection. Compasses are delicate instruments and should be cared for accordingly. Conduct a detailed inspection before you use your compass. The most important part is the floating dial. It must float freely. You must also ensure the sighting wire is straight, the glass encasement is not broken, and the numbers on the dial are readable.

(2) Effects Of Metal And Electricity. Metal objects and electrical sources can affect the performance of a compass (non-magnetic metals and alloys do not affect compass readings). To ensure the proper functioning of your compass, keep a safe distance from the following metal objects:

High-tension power lines	55 meters
Field gun, truck, or tank	10 meters
Telegraph or telephone wires and barbed wire	10 meters
Machine gun	2 meters
Rifle	1/2 meter
Steel rim glasses	1/3 meter

(3) Accuracy. A compass in good working condition is very accurate. However, you must periodically check your compass on a known line of direction, such as a surveyed azimuth using a declination station. If your compass has more than 3 degrees ± variation, do not use it.

(4) Protection. When traveling with the compass unfolded, make sure the rear sight is folded down onto the bezel ring. This will lock the floating dial and prevent vibration, as well as protect the crystal and rear sight from being damaged.

b. Techniques For Using The Compass. The lensatic compass is used to determine or follow magnetic azimuths during both the day and night. To use it with the maximum degree of accuracy, it is important that certain techniques be understood and properly applied. Like developing techniques for shooting a rifle, you must develop the proper holding position and practice until you master the techniques for accurately "shooting" an azimuth.

(1) Using The Centerhold Technique. This technique is faster and easier to use than the other techniques (see figure 1201-26).

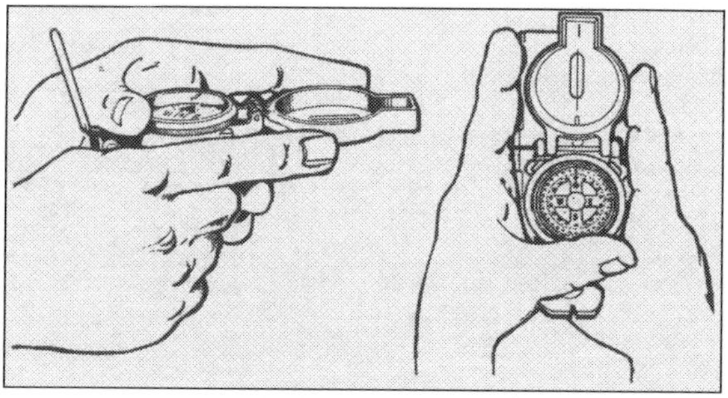

Figure 1201-26.

(a) Open the compass cover so that the cover forms a straightedge with the base.

(b) Move the rear sight to the rearmost position to allow the dial to float freely.

(c) Place your thumb through the thumb loop and form a steady base with your third and fourth fingers. Extend both index fingers along the sides of the compass.

(d) Place the thumb of the other hand between the lens (rear sight) and the bezel ring; place the remaining fingers around the fingers of the other hand.

(e) Pull your elbows firmly into your sides.

(f) Turn your entire body until the desired magnetic azimuth is under the fixed black index line.

(2) Using The Compass-To-Cheek Technique. This technique is used when you want to be more accurate (see figure 1201-27). Follow these steps.

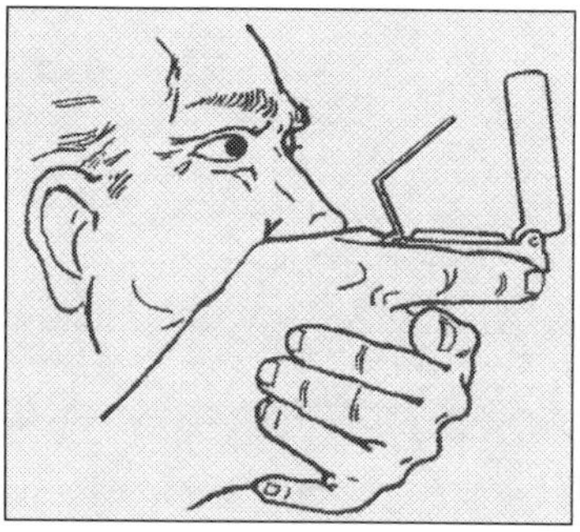

Figure 1201-27

(a) Open the compass so that the cover is vertical, forming a 90 degree angle with the base.

(b) Move the rear sight to the rearmost position to release the dial, then fold it slightly forward.

(c) Turn the thumb loop all the way down and insert your thumb. Form a loose fist under the compass, steady it with your other hand, and raise it to eye level.

(d) Look through the rear sight notch and center the front sighting wire in the rear sight notch.

(e) Keeping the compass level and the sights aligned, rotate your entire body until the sighting wire is lined up on a distant object.

(f) Glance down through the lens and read the azimuth directly under the black index line. The azimuth you read is the magnetic azimuth from your position to the distant object.

c. Converting Azimuths. Because there is an angular difference between grid north and magnetic north, a conversion from magnetic to grid or vice versa is needed.

(1) Declination Diagram. As previously discussed, azimuths measured with a protractor are grid azimuths (measured from grid north), and azimuths determined with the compass are magnetic azimuths (measured from magnetic north). You cannot follow a grid azimuth with a compass, nor can you plot a magnetic azimuth with a protractor because of the angular difference between grid north and magnetic north. This angular difference (between grid north and magnetic north) is called the G-M ANGLE (Grid-Magnetic angle). **The G-M angle varies for each map.** Because of this angular difference (the G-M angle), before you can plot a magnetic azimuth on a map, you must convert it to a grid azimuth. Likewise, before you can use a grid azimuth to navigate, you must convert it to a magnetic azimuth. Declination diagrams display the difference between grid and magnetic north. A complete set of instructions is included in MOST declination diagrams for your use in converting azimuths. The G-M angle often is not expressed as a whole degree, such as 1/2 degrees or 7 degrees 15'. Since you will not need to work with such precise numbers as minutes, round the G-M angle off to the nearest whole degree. If the G-M angle is 1/2o or 30' then round the angle up to the next highest whole degree.

(2) Conversion Notes. Refer to the conversion notes that appear with the declination diagrams explaining the use of the G-M angle (see figure 1201-28). One note provides instructions for converting a magnetic azimuth to a grid azimuth. The other provides instructions for converting a grid azimuth to a magnetic azimuth. The conversion (addition or subtraction) is governed by the direction of magnetic north relative to grid north.

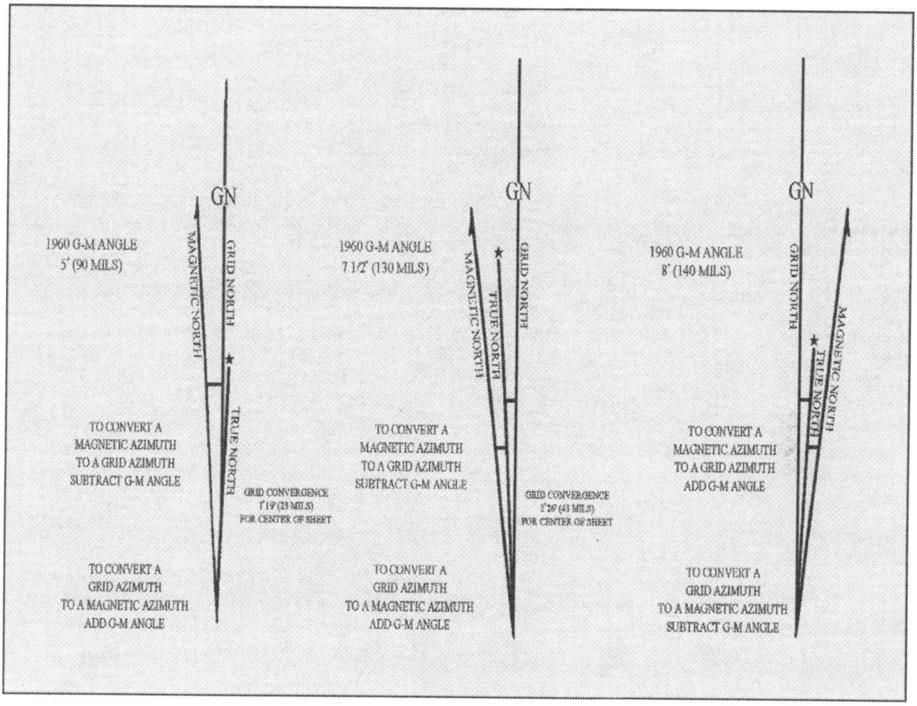

Figure 1201-28

(a) Converting Grid Azimuths To Magnetic Azimuths. The following is an example of part of a declination diagram:

TO CONVERT A
GRID AZIMUTH TO A
MAGNETIC AZIMUTH
SUBTRACT G-M ANGLE

On a map with a G-M angle of 15 degrees, to convert a 39 degrees grid azimuth to a magnetic azimuth you simply follow the instructions. Subtracting the G-M angle (15 degrees) from the grid azimuth (39 degrees), you get the correct magnetic azimuth of 24 degrees.

(b) Converting Magnetic Azimuths To Grid Azimuths. Again, examine the following declination diagram. The conversion note states:

TO CONVERT A
MAGNETIC AZIMUTH
TO A GRID AZIMUTH
ADD G-M ANGLE

To convert a 238 degrees magnetic azimuth to a grid azimuth you simply follow the instructions. Adding the G-M angle (15 degrees) to the magnetic azimuth (238 degrees), gives you the correct grid azimuth of 253 degrees.

d. Navigate During The Day. To successfully navigate, you must be able to determine from your map the best routes to follow, plot these routes, and follow your desired compass azimuth. Almost all routes plotted on a map are divided into a number of straight "legs." You must know the distance and azimuth for each leg of the march. A good navigator is able to stay on course by using the compass in conjunction with steering marks, and by understanding the factors that may cause you to wander off course.

(1) Determining Ground Distance By Pacing. Navigating involves many separate skills that are combined to successfully get you from one point to another. Pacing is another one of these skills. It helps you determine how much ground distance you have covered once you actually start moving on your route.

(2) Determining Your Average Pace Count. Many techniques have been developed for determining how many steps it takes the average Marine to walk a given distance. Predominant among these techniques is the "hundred meter pacing course." A distance of exactly 100 meters is marked on the ground. As you walk this distance, keep a count of the steps you take. Count every other step (every left foot) to determine your count. This is your "pace count" for the number of steps it takes you to walk 100 meters. The pace count for the average Marine is 60 paces (per 100 meters). If you cannot determine your pace count ahead of time, use this number.

(3) Now that you know your pace count, you need to know how to apply it to land navigation. Once you determine the distance that you need to cover (for example, how far you must travel on a leg of a patrol), you must convert this distance into the number of steps you must take to cover that distance. The formula to do this is simple:

$$\frac{D}{100} \times PC = P$$

D is the distance you must travel, PC is your pace count, and P is the number of paces you must take to travel that distance.

e. Selection And Use Of Steering Marks. Selecting steering marks is the last important thing you must know before you can start moving from one point to another. A steering mark is a well-defined object on your line of march on which you can guide. These objects can be natural or man-made (hill, tree, building, etc.), a celestial body (sun, stars, moon), or another person. One of the problems associated with selecting and using steering marks is that an object often looks good when you select it, but will become obscured as you approach it. This may confuse you and cause you to deviate from your intended line of march. The characteristics of good steering marks are discussed below and should be kept in mind when you select a steering mark.

(1) A good steering mark must have some distinct and unique features such as:

- Color
- Size
- Shape

A good steering mark will have all three. This assures you that it will continue to be recognizable as you approach it. A distant tree may appear to have a crooked limb which identifies it. However, upon entering the forest you may find dozens of widely separated trees with similar limbs. Your steering mark has then served little purpose. A steering mark must have a feature that distinguishes it from all other similar objects on your line of march.

(2) If several easily distinguishable objects appear along your line of march, the best steering mark would be the most distant object and/or the highest. This will enable you to travel farther with fewer references to the compass. The higher steering mark is not as easily lost to sight as is a low steering mark that may blend into the background as you approach it.

(3) A steering mark should be continuously visible. If the terrain or vegetation ever blocks the steering mark from view, take out your compass and select an intermediate steering mark. Continue using intermediate

steering marks until your original steering mark comes back into view. To enter a wooded area thinking that you will come out on the other side still heading in the correct direction is foolish. In thick vegetation, you can quickly become disoriented and emerge from the thicket heading the wrong way with no steering mark in view.

(4) If you are navigating alone and there are no usable steering marks to your front, it is still possible to proceed on an azimuth by referring to a back steering mark. Simply determine the back azimuth of the azimuth you are following. Now face about, and see if there is a prominent object on line with the back azimuth. If none exist, erect one where you are located (place a stick in the ground, pile up some rocks, etc.) When you continue on your route, use the same principles that apply for a forward steering mark, except that you use back azimuths to maintain the intended line of march and you must check your compass more frequently.

(5) If appropriate landmarks are not available at night, you may select a plainly visible star along your line of march to serve as a steering mark. Remember though, due to the Earth's rotation, any star that you choose will eventually either disappear under the horizon or will move too high in the sky to be of further use. If this happens, choose another steering mark.

(6) If no natural or man-made object is visible on your line of march, have another Marine move forward to serve as a steering mark. The distance he moves is determined by the terrain, amount of visibility, and the tactical situation. Since this is time-consuming and may be tactically compromising, you should never use this technique except as a last resort.

(7) Steering marks are selected as the march progresses. Shoot your azimuth, select the best steering mark on this azimuth, and head to it. Whether you are navigating during daylight or periods of reduced visibility, through densely wooded areas or open terrain, over short or great distances, every step you take should be toward a selected steering mark. The natural human tendency to veer off course, even on short distances, is too great to needlessly trust your ability to walk a straight line. So, whether it is an object which meets all the requirements of an ideal steering mark, a star, or something as insignificant as a tiny patch of light in the foliage, ALWAYS, choose something on your line of march.

f. Following A Compass Azimuth During The Day. The procedure for following a compass azimuth during the day utilizes the compass-to-cheek technique and the selection and use of appropriate steering marks. Let's begin.

(1) Shoot your desired azimuth using the compass-to-cheek technique.

(2) Using the sighting slot and the sighting wire, choose the best steering mark that is directly in line with your azimuth.

(3) Recheck your azimuth. The few seconds required for this are well spent.

(4) Close your compass to protect it during movement and step off towards your steering mark.

(5) Periodically spot check your azimuth by using the centerhold technique. You can do this without stopping.

(6) When you arrive at your steering mark, stop and select a new steering mark. Continue repeating the steps until your journey is complete.

NOTE. If your steering mark becomes hidden from your view, stop and select another steering mark immediately.

g. Navigating Using Dead Reckoning. The term "dead reckoning" derives from the seafaring days before the arrival of modern navigational equipment when mariners would "deduce" (ded) how far they had traveled on their course. Based on this deduction, they would "reckon" where they were located. To save room in the ship's log, entries were abbreviated to read "ded-reckon." "Dead-reckoning" is a means of navigating using only pace counts and azimuths. Terrain association is not used. In thick jungles, barren terrain, or when visibility is limited, you will not be able to associate your current position with visible terrain features on the ground. Dead reckoning may be the only means you have available of knowing where you are located on your map. To navigate by dead reckoning, you must accomplish the following:

(1) Determine Location, Distance And Azimuth. Determine your own location by any means. Then determine the distance and magnetic azimuth between each leg of your journey.

(2) Step Off. During movement, make frequent reference to your compass to ensure you are moving in the correct direction. Carefully select your steering marks (navigating with steering marks is NOT the same as navigating by terrain association as steering marks are rarely shown on a map). Maintain an accurate pace count. When you have paced off the appropriate number of steps on the appropriate azimuth, then you have reached your objective (or checkpoint). Repeat the steps until you reach your final destination.

h. Following a Compass Azimuth at Night. The procedure for following an azimuth at night is basically the same as for following an azimuth during daylight. In fact, on a bright, moonlit night, the procedures are identical since you can see the compass dial clearly. However, because of the reduced visibility, night navigation will require further consideration in selecting steering marks and applying the features of the lensatic compass intended for night use.

(1) Setting The Compass For Night Use. The lensatic compass is equipped with special features which enable you to follow an azimuth during periods of darkness. Around the base of the bezel ring is a series of 120 notches. On the forward edge of the body of the compass is a tiny bezel detent spring with its tip seated in one of the notches. As the bezel ring is turned, the spring moves from notch-to-notch producing "clicks." Each click equals 3°. This clicking action, used in conjunction with the luminous markings on the compass, provides a means of setting your compass at night.

(2) Using The Click Technique. At night, an azimuth can be set on the compass by using the click technique (see figure 1201-29). Remember that the bezel ring contains 120 clicks; each click equals 3 degrees. Use the following steps to set a compass for night navigation.

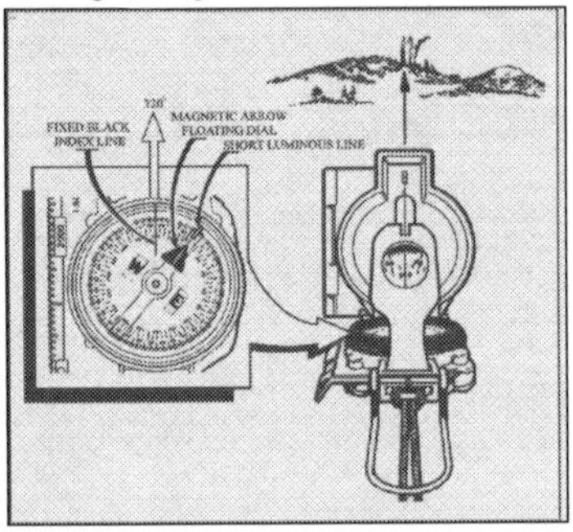

Figure 1201-29

(a) Rotate the bezel ring until the short luminous line is directly over the fixed black index line.

(b) Divide the azimuth you plan to travel by 3 to get the number of clicks you must rotate the bezel ring.

(c) Rotate the bezel ring the desired number of clicks in a counterclockwise direction.

NOTE: If the number of clicks is greater than 60, you can subtract this from 120 and rotate the bezel ring in a clockwise direction. For example, you would move the bezel ring 20 clicks clockwise instead of 100 clicks counterclockwise.

(d) Hold the compass in the centerhold position and rotate your body until the north seeking arrow is under the short luminous line.

(e) Step off by following the line indicated by the two luminous sighting dots on the compass cover.

NOTE: Do not follow the north seeking arrow.

i. Selecting Steering Marks For Night Use.

(1) The general characteristics of a good steering mark for daylight navigation apply to steering marks selected for night navigation. However, as darkness approaches, keep the following factors in mind when selecting steering marks:

(a) Colors disappear at night and objects appear as black or gray silhouettes. If you select a steering mark during daylight because of its distinctive color, it will be of little value as darkness sets in.

(b) During darkness, steering marks must be closer than during daylight.

(c) During darkness, steering marks must have a distinctive silhouette. The silhouette of trees, bushes, and similar objects will change because you see them from slightly different angles as you move up or down hills and when you are detouring obstacles. This applies during daylight also, but to a lesser degree.

(2) Following An Azimuth Without Steering Marks. At times it will be so dark that you can see only a few feet in front of you. When this happens, you have to follow your azimuth without steering marks. The only way that you can stay on your azimuth is by frequently looking at the compass. Practice, experience, and patience will enable you to successfully follow your azimuth under such conditions. But the human tendency to drift will make this technique less accurate than if you use steering marks. You will have to overcome this inaccuracy by thoughtfully selecting your route and checkpoints.

j. Execution Of Navigation At Night. The techniques which produce successful navigation at night are basically the same as those employed during daylight. However, special consideration and emphasis must be given to those aspects of navigation which are affected by the limitations imposed by darkness.

(1) Terrain Association. Just as in day land navigation, at night you should be familiar with the terrain that you will be negotiating. If you should be going down hill when you start off and you are going up hill, then something is wrong. You won't be able to use terrain association to the same extent as during the day, so you may have to rely on dead reckoning. Regardless of the situation, maintain an accurate azimuth and pace count.

(2) Bypassing Large Obstacles At Night. Bypassing an unexpected obstacle at night is fairly simple--just use the 90 degree offset technique. To make a 90 degree turn to the right, hold the compass using the centerhold technique; turn until the center of the luminous letter E is under the luminous line (do NOT move the bezel ring). To make a 90 degree turn to the left, turn until the center of the luminous letter W is under the luminous line. This does not require changing the compass setting (bezel ring), and it ensures accurate 90 degree turns.

(3) NEVER use uncovered flashlights during night navigation because it will reduce your "night vision" and possibly disclose your location to the enemy. If a light must be used, it should be a filtered light and should ALWAYS be used under a poncho or similar cover. If your flashlight does not have a filter, cover the lens with a thin layer of mud so that it reveals only enough light to serve your purpose. Many Marines of all ranks hold the mistaken belief that because they have a red lens on their flashlight, they can not be seen by the enemy. THEY ARE WRONG! They can be seen HUNDREDS of meters away.

(4) Checkpoints. Checkpoints must be easy to recognize. You should be familiar with the characteristics of the checkpoint itself and the terrain features in its vicinity. You should know when you are in the vicinity of a checkpoint. Although not always possible, a checkpoint for night navigation should be of such a nature that missing it entirely is nearly impossible even though you are slightly off course. For this reason a line checkpoint is much superior to a point checkpoint for navigating at night.

9. DETERMINE YOUR LOCATION.

a. Orient A Map. Your first step when navigating in the field is to orient your map. A map is oriented when its north and south correspond to the north and south on the ground. There are two ways of orienting your map - through the use of a compass and by terrain association.

(1) Using A Compass. When orienting a map with a compass, remember that compasses measure magnetic azimuths. Since the north-seeking arrow of the compass points to magnetic north, pay special attention to the declination diagram. Use the following technique to orient your map:

(a) With the map flat on the ground, place the straightedge (on the left side of the compass) along the magnetic north arrow on the declination diagram so that the cover of the compass is pointing toward the top of the map. This will put the fixed black index line of the compass parallel to the magnetic north arrow of the declination diagram (see figure 1201-30).

(b) Keeping the compass aligned as directed above, rotate the map and compass simultaneously until the north-seeking arrow is below the fixed black index line on the compass. Your map is now oriented.

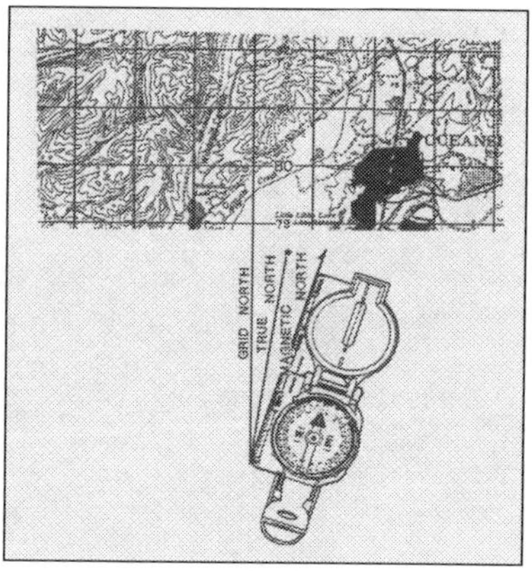

Figure 1201-30

(2) Using Terrain Association. You can orient your map using terrain association when a compass is not available or when you have to make quick references as you move across country. Using this technique requires careful examination of the map and the features on the ground.

(a) Identify Terrain Features. Identify prominent terrain features on the map that you can find on the ground.

(b) Align Terrain Features With The Map. If there is a tower to your right front, then orient the map so that the tower is to your right front. If there is a road off to your left, then ensure the road on the map is parallel to the road on the ground. Once all of the features are lined up, your map is oriented.

b. Determining Your Position. Now that you can orient your map, you must find your location on a map. If you know your approximate location on a map, a study of nearby terrain features will help you determine your position.

(1) Determining Your Location By Inspection. You are standing in the vicinity of several prominent features which can easily be located on the map. By orienting the map and estimating your relation to these features, you should have no difficulty in determining your location.

(2) Determining Your Location By One-Point Resection. One-point resection is an accurate technique of determining your location when you are on or near a linear feature that you can identify both on the ground and on the map. You must also be able to identify another prominent feature, both on ground and on the map. To determine your location by one-point resection follow these steps:

(a) Identify the linear terrain feature that you are located on or near in respect to the ground on your map.

(b) Identify a prominent feature on the ground and locate that feature on your map.

(c) Using the compass-to-cheek technique, sight in on the feature and read the magnetic azimuth.

(d) Convert the magnetic azimuth to a grid azimuth.

(e) Convert this grid azimuth to a grid back azimuth.

(f) With your protractor, plot this grid back azimuth from the feature on the map and extend it until it crosses the linear feature.

(g) Conduct a map inspection to verify your resection.

When selecting a terrain feature, choose one that is perpendicular to the axis of the linear terrain feature so that when you plot the back azimuth on the map, the line will cross the linear feature more or less at a right angle. Figure 1201-31 illustrates the result of a good selection and a poor selection.

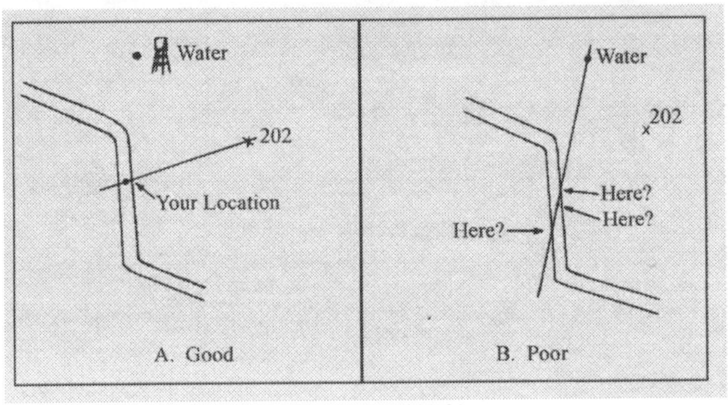

Figure 1201-31

NOTE: A one-point resection can still be done if you are not along a linear feature. Merely estimate the distance between you and your prominent terrain feature. Plot this distance along the back azimuth to get your approximate location.

(3) Determining Your Location By Two-Point Resection. Usually you will find that you are not located on or near a prominent linear feature. Since the accuracy of a one-point resection under these conditions depends on your ability to accurately estimate distance, it is better to use a two-point resection. The procedures for two-point resections are basically the same as for one-point resections except you must select two features instead of one. The back azimuths from each feature is determined and plotted on your map. You are located at the point where these lines cross. If you have a compass and a protractor then follow these steps.

(a) Select two prominent features on the ground whose positions can be located on the map. These features should be at least 30° but not greater that 150° apart (see figure 1201-32).

(b) Using the compass-to-cheek technique, determine the magnetic azimuth to each object.

(c) Convert these magnetic azimuths to grid back azimuths.

(d) With your protractor, draw the respective back azimuths from these two points on your map.

(e) Extend the azimuth lines from these two points until they intersect. You are located at the point where these two lines cross.

(f) Conduct a map inspection to verify your position.

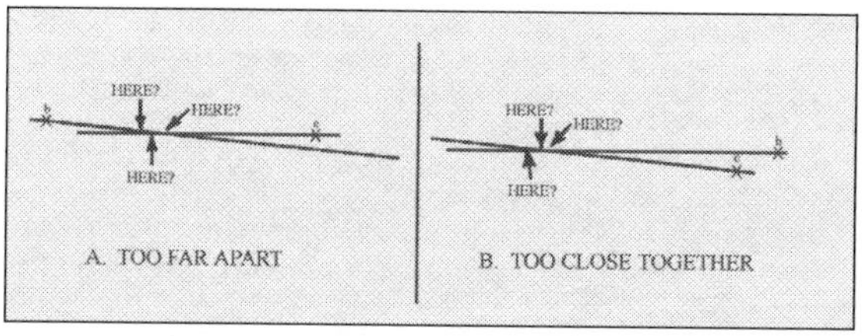

Figure 1201-32

10. BY-PASSING OBSTACLES. Despite the amount of care you take in determining routes, choosing steering marks, and using your compass to maintain a straight line of march, unexpected obstacles will often be encountered which will force you to detour from your intended line of march. There are several ways to bypass obstacles.

a. Detouring Small Obstacles. If you encounter small obstacles, such as trees and bushes, go completely around the obstacle to the center of the opposite side and continue your movement.

b. Detouring Large Obstacles Using Steering Marks. A large obstacle, such as a pond or impassable swamp, can be successfully bypassed if there is a clearly defined steering mark on the far side of the obstacle. Decide carefully whether the object will serve as a suitable steering mark. Remember that an object may have a distinguishing characteristic when observed from a distance, but lose its identity as it is approached. As a final precaution, always leave some mark, such as a small pile of rocks or a blaze (distinctive mark) on a tree on your side of the obstacle, so you can return if you become confused while bypassing the obstacle.

EXAMPLE: You have been following an azimuth of 212 degrees toward your next checkpoint when you come upon an impassable swamp. On the far side of the swamp you see a large pile of rocks that is on a 212 degree azimuth (your original line of march). You can see that this is the only pile of rocks in the area. (If there were others, you would have to carefully study the characteristics of each rock pile and select several features, such as size, color, and nearby objects, which would easily distinguish it from the other rock piles.) Estimate the distance across the swamp, leave a mark on your side of the swamp, put your compass away, and proceed to the rock pile by the easiest route. Upon reaching it you may continue toward your checkpoint along a 212 degree azimuth. (Remember to deduct the estimated distance across the swamp from the remaining distance on that leg.)

c. Detouring Large Obstacles Using Back Steering Marks. If there is no suitable steering marks on the far side of the obstacle, you can bypass the obstacle if there is an easily distinguishable object on your side which can be used as a back steering mark. If there isn't a suitable natural steering mark, then make one. Carefully examine the terrain and shoreline along your line of march on the far side of the obstacle (this will help you estimate when you are at the correct position). Now, estimate the distance across the pond, put away your compass, and proceed to the other side by the easiest route. When you think you are back on your line of march, look back across the obstacle for your back steering mark. Now break out your compass and shoot an azimuth to the back steering mark. If you are in the right place, the mark will lie on the back azimuth of your original azimuth. If it is not, you have not returned to your original line of march. You must move to your right or left until you find the spot from which the mark lies at the correct back azimuth. When you have found this spot, you have returned to your original line of march and may resume your original azimuth. (Again, don't forget to deduct the estimated distance across the pond from your remaining distance on that leg.)

d. Detouring Large Obstacles Using The 90 Degree Off-Set Technique. If you are navigating at night or there is absolutely no way to use steering marks, you can still stay on track using this technique.

(1) Detour the obstacle by moving at right angles (plus or minus 90 degrees) for specified distances. For example (see figure 1201-33), while moving on an azimuth of 90 degrees, change your azimuth to 180 degrees and travel for 100 meters; change your azimuth to 90 degrees and travel for 150 meters; change your azimuth to 360 degrees and travel for 100 meters; then change your azimuth to 90 degrees and you are back on your original azimuth line, 150 meters closer to your next checkpoint.

(2) You must maintain an accurate pace count for each leg of your detour. All paces which are in the direction of your original line of march must be combined so that, upon returning to your original line of march, you know the distances yet to be traveled to reach your next checkpoint. All paces at right angles (90 degrees) to your original line of march must be computed so that, upon passing the obstacle, you know how many paces it will take to return you to your original line of march. You should return to the original line of march as soon as possible after passing the obstacle. The only sure way of knowing that you are back on this line is by maintaining an accurate pace count.

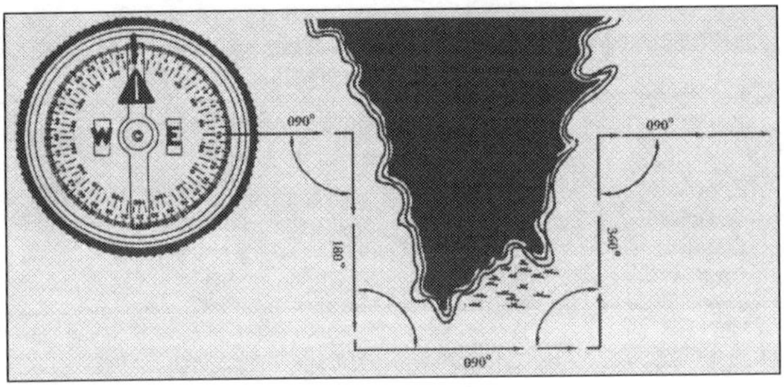

Figure 1201-33

APPENDIXES:

A. Example of Land Navigation Evaluation Checklist
B. Sample of Practical Application

REFERENCE:

1. FMFM 21-26, Land Navigation

APPENDIX A

SAMPLE PE CHECKLIST

LAND NAVIGATION PERFORMANCE EVALUATION CHECKLIST

Learning Objectives.

1. Without the aid of references and given a map, protractor, and eight (8) digit grid coordinates, plot the points on the map in order to reinforce your knowledge and skills in navigating on land, per FM21-26. (SCRS12.01.01b)

2. Without the aid of references and given a map, protractor, and eight (8) digit grid coordinates, measure the distance from one point to another in order to reinforce your knowledge and skills in navigating on land, per FM21-26. (SCRS12.01.01c)

3. Without the aid of references and given a map, protractor, and eight (8) digit grid coordinates, determine the grid azimuth between points in order to reinforce your knowledge and skills in navigating on land, per FM21-26. (SCRS12.01.01d)

4. With the aid of references and given a checklist, area of operations, lensatic compass, map of the area, eight (8) digit grid coordinates, starting point, finishing point, and as a member of a two (2) man team, navigate during daylight hours in order to locate five (5) checkpoints on a map of the area in a time limit of four (4) hours in order to reinforce your knowledge and skills in navigating on land, per FM21-26. (SCRS12.01.01h)

5. With the aid of references and given a checklist, area of operations, lensatic compass, map of the area, eight (8) digit grid coordinates, starting point, finishing point, and as a member of a two (2) man team, navigate during darkness hours on order to locate three (3) checkpoints on a map of the area in a time limit of four (4) hours in order to reinforce your knowledge and skills in navigating on land, per FM21-26. (SCRS12.01.01I)

Instructions To The Student.

CLASSROOM PORTION. "For this class, you will be required to plot a total of (10) eight digit grid coordinates, measure the distance between each point, determine the grid azimuth between the points, and convert the grid azimuth to a magnetic azimuth. You will be graded by a checklist in accordance with FM 21-26. You will have 30 minutes to complete this task. If you have questions feel free to ask me, your time begins now.... Good luck."

FIELD PORTION. "For the day land navigation, I will show you what the start-points and checkpoints look like. Next we will move over to the 100 meter pace course, so you can determine your pace count for 100 meters. At this time listen up for your safety brief, Instructor Note: (in accordance with your Academy SOP). You will have 4 hours to locate five points. The time on deck is now _____ , your drop dead time is ______. Your drop dead time means, the time that you will be back here inside the CP. Your escape azimuth is ____________. Good Luck!!"

"For the night land navigation, we will move over to the 100 meter pace course, so you can determine your pace count for 100 meters at night. At this time listen up for your safety brief, Instructor Note: (in accordance with your Academy SOP). You will have 4 hours to locate three points. The time on deck is now _____ , your drop dead time is ______. Your drop dead time means, the time that you will be back here inside the CP. Your escape azimuth is ____________. Good Luck!!"

Land Navigation
Performance Evaluation Checklist

CARD # ________

NAME: ______________________________ PLATOON: ____________________

NAME: ______________________________ GM ANGLE IS __________ DEGREES

DAY STARTING POINT # ______

STARTING POINT: ________________ (plotted Correctly **1 point**) ______

Plot Point	Grid	**1 Point Each** Plotted Correctly	**1 Point Each** Grid Azimuth	Magnetic Azimuth	**1 Point Each** Distance	Total Points
#1						
#2						
#3						
#4						
#5						
					Total Points	

NIGHT STARTING POINT # ______

STARTING POINT: ________________ (plotted Correctly **1 point**) ______

Plot Point	Grid	**1 Point Each** Plotted Correctly	**1 Point Each** Grid Azimuth	Magnetic Azimuth	**1 Point Each** Distance	Total Points
#1						
#2						
#3						
					Total Points	

Day Point Identification Answers	**8 Points Each**
Point # 1:	
Point # 2:	
Point # 3:	
Point # 4:	
Point # 5:	
Total Points	

Night Point Identification Answers	**8 Points Each**
Point # 1:	
Point # 2:	
Point # 3:	
Total Points	

Day Time: ______ Time Points (0 or 5) ________ Night Time: ______ Time Points (0 or 5) ________

Total Points Land Nav: _______ (Final Grade)

Student's Signature: ______________________________

Evaluator's Signature: ______________________________

APPENDIX B

SERGEANTS COURSE
SAMPLE PRACTICAL APPLICATION

1. Locate and identify the most prominent terrain or man made feature for the following grid location.

    a. 93326721____________________

    b. 89927653____________________

    c. 91556780____________________

    d. 92567314____________________

    e. 89307351____________________

2. Determine the straight-line distance between the following points.

    a. Garfield High School (9979) and Bel Air (9879).
    ___________ meters

    b. The intersection of route 640 and 234 (9078) and Bench Mark 101.9 (9177).
    ___________ meters

    c. Greenwood Church (9578) and Neabsco Church (9876).
    ___________ meters

3. Determine the irregular map distance between the following points.

    a. The intersection of route 640 and route 234 (9078) and the intersection of route 646 and route 234 (8878) along route 234.
    ___________ meters

    b. Bench Mark 124.0 (8779) and Lookout Tower (8781).
    ___________ meters

    c. The intersection of interstate 95 and route 610 (9975) and intersection of interstate 95 and route 234 (9772) along interstate 95.
    ___________ meters

4. Plot the following azimuths with your protractor and answer the questions in the space provided below.

    a. From Garfield High School (9979), plot a grid azimuth of 91 degrees. What topographical feature is located 2000 meters away along this azimuth?

    ____

    b. From LZ Duck (8274), plot a gird azimuth of 232 degrees. What topographical feature is located 5440 meters away along this azimuth?

    ____

    c. From Cole School (8881), plot a grid azimuth of 195 degrees. What topographical feature is located 2050 meters away along this azimuth?

____

5. Resection (8 digit grid is required)

   a. You are moving NNE along an unimproved road located in grid 7078. You take a azimuth of the lake toward the East of 105 degrees (magnetic), and another of the lake toward the South of 211 degrees (magnetic). What is your position? ______________

   b. You are located somewhere 9167, you take an azimuth of 30 degrees (magnetic) to a dam that is 610 meters away. What is your located? ______________

   c. You are moving South along an unimproved road located in grid 9667. You take an azimuth SSW to Ashrust School, which yields 229 degrees (magnetic), and another to the SE to a water tower which yields 146 degrees (magnetic). What is your position? ______________

Made in the USA
Monee, IL
05 November 2026

67808326R00105